Motorways Made Easy

The M40 passes through the lovely countryside of the Chilterns near Lewknor in Oxfordshire.

Motorways Made Easy

Editor Penny Hicks
Designer M A Preedy MSIAD
Gazetteer compiled by
Daphne Jolley

Maps produced by the
Cartographic Department
of the Automobile Association.
The illustrations on pages 4, 6 and 7
are reproduced from The AA
Junior Atlas by kind permission of
Hamish Hamilton.

The contents of this book are
believed correct at the time of
printing. Nevertheless, the
publisher can accept no responsi-
bility for errors or omissions or
changes in the details given.

ISBN 0 86 145 220 8

Typeset by: Vantage Photosetting
Co. Ltd, Eastleigh, Hants.

Printed and bound by:
Graficromo SA, Spain

Published by the Automobile
Association, Fanum House,
Basingstoke, Hampshire RG21 2EA.

Tramway Museum, Crich, Derbyshire (M1)

In contrast to the busy M6 around Carlisle, little traffic is encountered on the quiet roads of the Kielder Forest
twenty-six miles north-east on the Scottish Borders.

Contents

Arley Hall and gardens (M6)

Making the Motorway Easy

The first 72 miles of the Motorway from London to the Midlands was opened in 1959 amidst a blaze of publicity which heralded 'a new era in motoring'. (In fact, the very first section of motorway to open was the 8-mile Preston bypass, now part of the M6, which opened almost a year earlier.) A master plan for the nation's road system, first mooted before the war, was at last becoming a reality. Some 25 years later it still isn't finished, but we do have over 1500 miles of motorway in use, with a complete system of over 2000 miles planned. Not the least achievement will be the long-awaited M25 London Orbital motorway which should completely encircle the capital by 1986.

Motorways are now a part of every driver's life, but many have mixed feelings about them. The

Finding your way
First of all we include six pages of Route Planning Maps showing motorways and other main roads in an easily manageable scale for planning a long journey. In the main body of the book, we take relatively short stretches of motorway, rarely more than 50 miles at a time, and detail them in full. Each map page contains a Thermo-chart indicating junctions, direction signs at junctions and distances between them. The locations of any service areas are also shown. Next to the chart a map shows the same piece of motorway in relation to its surroundings with other roads, towns, villages etc. Some motorway junctions are less straightforward to negotiate than others and detailed, large-scale diagrams showing their layouts are also in-

beyond all recognition. Rather than being the epitome of bad catering, they now offer a creditable three-course meal often with a glass of alcohol-free wine or beer. Service and cleanliness have improved too. The advantage of the service areas is that they are convenient and usually always open – where else can you get a drink and a snack at 3am? As well as their service station, toilet and restaurant facilities they may offer a fast food take-away, a shop, a picnic-area and somewhere for the children to amuse themselves for a while. Facilities available at each service area are included in the book.

Although most travellers use the service areas, there are many who would prefer to ven-

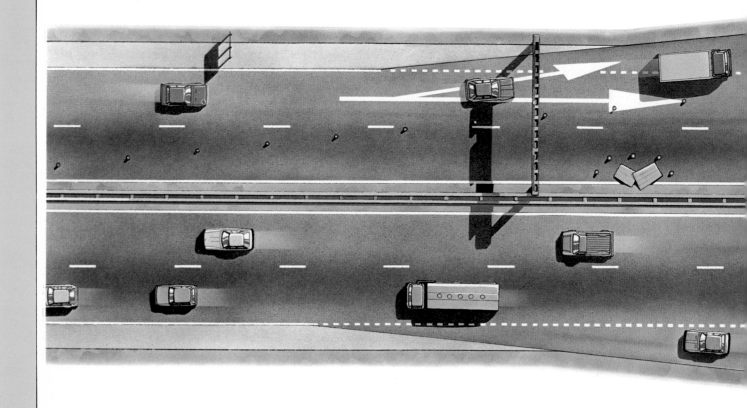

speed and ease of covering long distances is greatly appreciated, and statistics prove that motorways are six times safer than other roads. Nevertheless, motorway driving can be tedious in the extreme, following a seemingly endless ribbon of highway into the distance with little of interest on either side. Drivers have fewer road hazards and features to keep them alert and passengers have even less to alleviate the boredom. This book is designed to help everyone get the most from their motorway journey.

cluded. Facing each map is a page giving essential information about motorway services, places of interest and listing hotels, restaurants and garages near to each junction.

Breaking the journey
The motorway system is served by over 40 service areas which in the past gained a reputation for poor food and expensive petrol. Thankfully, loudly voiced public opinion won the day and in general the service areas have improved

ture off the motorway for a break and this doesn't necessarily mean adding great distances to the journey. Accompanying each map page are lists of alternative, AA-recommended places to eat and garages within a short distance of motorway junctions, generally listed in order of their proximity to the motorway junction. Not too far to go, but far enough to bring a welcome change of pace and a greater variety. For anyone wishing to break their journey overnight, a

Advised maximum speed

Leave motorway at next exit

Lane clear

list of AA-appointed hotels is also included and, naturally, it is also possible to obtain refreshment there too. The list covers not only hotels but also extends to the less expensive private hotels, guesthouses, inns and, occasionally, farmhouses. Many people find that these latter categories provide a more friendly and informal welcome for an overnight stay. Names, addresses and telephone numbers are provided in this book, but more detailed information can be found in the AA *Members' Handbook* or in one of the AA Superguides: *Hotels and Restaurants in Britain*; *Guesthouses, Farmhouses and Inns in*

easy reach of the motorways are included in this book, indicated 'PS' in the listings. A full list of picnic sites can be found in the AA Superguide *Camping and Caravanning in Britain*.

The main criticism of motorway travelling is undoubtedly the dullness of it, so to add a bit of interest on each page we describe the general character of the countryside through which the section of motorway passes. We also list a selection of places of interest and other attractions in close proximity to the junctions.

information, we include details of local radio wavelengths for up-to-date information as you travel. A map showing all the radio stations and the areas covered by them can be found on pages 22–23, while each motorway section page includes the relevant wavelengths for that area.

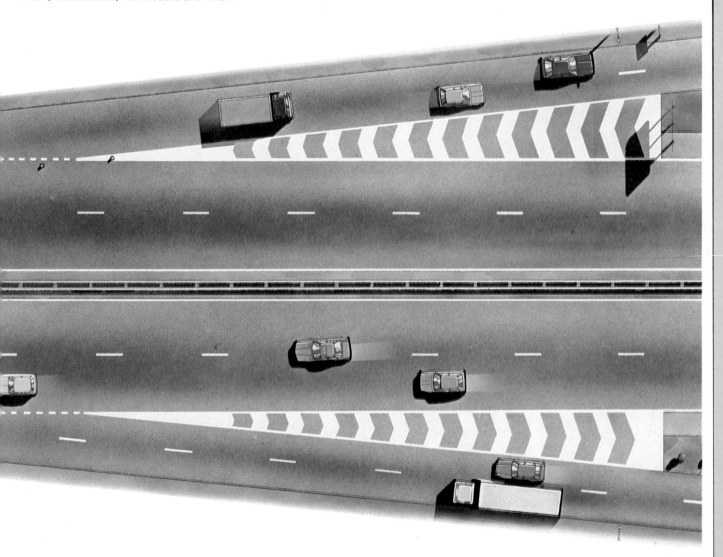

Britain; Bistros, Inns and Wine Bars in Britain.

On a fine day it is a nice idea to break the journey with a picnic and to relax for an hour in the fresh air away from the noise and fumes. The AA inspects and recommends hundreds of picnic sites all over Britain where the facilities are of a good standard and those which are within

Countdown markers: 300yds, 200yds and 100yds before exit slip roads.

Practical help

Motorway driving is quite different to that on other roads: techniques have to be adapted accordingly and it is essential to prepare the car for a high-speed, long distance journey so we include a special feature dealing with this. The motorway system itself is described, together with the basic regulations and all the traffic signals. We also explain how different weather conditions can affect the motorways and driving on them, pointing out some of the particular trouble spots to avoid.

It is also helpful to know what is going on ahead of you, whether it be road works or blockages caused by accidents or bad weather. As well as listing telephone numbers for advance

On the face of it, a journey by motorway would seem to require a minimum of planning and it is true that you might just travel from one end to the other without a thought. The wide and comprehensive range of information given in this book, however, offers everyone who uses motorways greater freedom and flexibility. The journey can be made so much more pleasant by taking a break from time to time, either in one of the service areas or at one of the hundreds of other stops mentioned in this book. Driver and passengers alike will be grateful when a potentially dull journey turns into a pleasant trip.

AA-Appointed

How the AA inspects and classifies its thousands of appointed establishments.

Hotels

The AA has been inspecting hotels for some 75 years, during which time it has gained invaluable experience as well as making some contribution to the high standards in British hotels – a fact acknowledged by the hotel trade itself. To ensure the full co-operation of hotel proprietors, the request for an inspection must come from them first. Following such a request, one of our team of experienced inspectors will book into the hotel, posing as a member of the general public. It is only after the bill is paid at the end of the stay that the inspector will reveal his or her true identity and ask to be shown around the rest of the hotel. A thorough report is then sent to the Head Office Appointments Committee who make the final decision on listing and classification. The famous star-ratings (black, white or red – see below) are awarded purely on the facilities which are offered by the hotels:

Black stars denote hotels offering traditional service in traditional accommodation. The majority of AA-appointed hotels are in this category.

★ Good hotels and inns, generally of small scale and with acceptable facilities and furnishings.

★★ Hotels offering a higher standard of accommodation, with some private bathrooms/showers; lavatories on all floors; wider choice of food.

★★★ Well-appointed hotels; a good proportion of bedrooms with private bathrooms/showers.

★★★★ Exceptionally well-appointed hotels offering high standards of comfort and service, the majority of bedrooms having private bathrooms/showers.

★★★★★ Luxury hotels offering the highest international standards.

Note: Hotels often satisfy some of the requirements for a higher classification than that awarded. In provincial 5-star hotels some of the services are provided on a more informal and restricted basis.

☆ White stars indicate establishments high in amenities but which have deliberately limited services, designed and operated to cater mainly for short-stay guests.

⚤ Denotes an AA Country House hotel, where a relaxed, informal atmosphere prevails. Some of the facilities may differ from those at urban hotels of the same classification.

○ Hotels due to open during the currency of this book which had not been inspected at the time of going to press.

If an inspector considers a hotel to be of a quality far above that implied by the star rating, it will be recommended for a special Red Star award – a much-coveted accolade held by only 51 hotels at present. In addition there are three special merit awards for particular aspects of the hotel's accommodation and service:

H Hospitality, friendliness and service well above the average for hotels similarly classified.

B Bedrooms significantly better than those to be expected within the star classification.

L Lounges, bars and public areas significantly above the standard implied by the star classification.

If the cuisine is exceptional, rosettes are awarded – see under 'Restaurants' for a full explanation.

Guesthouses

These are different from, but not necessarily inferior to, AA-appointed hotels, offering an alternative for those who prefer inexpensive and not too elaborate accommodation. Having said that, some guesthouses are quite luxurious, with en-suite bathrooms, colour TVs in bedrooms and very elegant public rooms. Generally speaking, though, this category of accommodation provides clean, comfortable and homely surroundings. None of them will be fully licensed, but many will have a residential or restaurant licence and some may only offer bed and breakfast. Look for 'GH' in the list of places to stay.

Farmhouses

A recent survey undertaken by the AA has found a very high level of satisfaction with both the food and the accommodation offered by farmhouses in Britain. Most of those we list are working farms, so children should be very closely supervised, but the proprietors will always be pleased to advise where they can safely go around the farm. As few as two letting bedrooms are acceptable for AA recommendation, making a stay on a farm particularly cosy. They are indicated 'FH' in the list of places to stay.

Inns

There is a very long tradition in this country of inns offering food and shelter to weary travellers, many of whom still seek out a welcoming hostelry in preference to any other kind of accommodation. In actual fact, the keeper of an inn is under a legal obligation to provide food, drink (not necessarily alcoholic) and accommodation to any bona-fide traveller. Inns included in this book will have between 3 and 15 letting bedrooms and a suitable room for taking breakfast.

Restaurants

Restaurants are assessed differently from hotels in that it is usually the AA inspector who seeks out the establishment, but they are still visited anonymously and judged on their cuisine, at-mosphere, service and value for money. A report is submitted to the Restaurant Committee who will award 'crossed knives and forks' ratings, or in exceptional cases rosettes (see below) to denote the amenities:

✕ Modest but good restaurant

✕✕ Restaurant offering a higher standard of comfort than above

✕✕✕ Well-appointed restaurant

✕✕✕✕ Exceptionally well-appointed restaurant

✕✕✕✕✕ Luxury restaurant

Rosettes

Rosettes are awarded where our inspectors consider the food to be of a particularly high standard:

❀ Hotel or restaurant where the cuisine is considered to be of a higher standard than is expected in an establishment within its classification.

❀❀ Hotel or restaurant offering very much above-the-average food irrespective of classification.

❀❀❀ Hotel or restaurant offering outstanding food, irrespective of classification.

Because of the vast range of restaurants in the country, the AA decided a few years ago that it should look at those which may not qualify for 'knife and fork' rating but which offer good food at reasonable prices. Published in the AA Super-guide *Bistros, Inns and Wine Bars in Britain,* these establishments include pubs, bistros, wine bars, Italian, Chinese and Indian restaurants, and small country cafés. You will find them listed without classification under the 'Where to Eat' sections, and at any one of them you should be able to get a good two-course meal for around £5 or less – considerably less in some cases.

Garages

All of the garages which we list in this book have been regularly inspected by the AA's engineers to ensure that requisite standards are maintained. Full details of the facilities available can be found in the AA *Members' Handbook*, but in this book we give the name and address together with any special franchise that each garage may hold. These are abbreviated in the text as shown below:

Cars							
AC	AC	Frd	Ford	RR	Rolls-Royce/	BSA	BSA
AM	Aston Martin	FSO	FSO		Bentley	CZ	CZ
AR	Alfa Romeo	Hon	Honda	Sab	Saab	Duc	Ducati
ARO	ARO	Jep	Jeep	Sko	Skoda	Gar	Garelli
Aud	Audi/NSU	Lad	Lada	Sub	Subaru	Gil	Gilera
BL	Austin/Rover/	Lnc	Lancia	Suz	Suzuki	Hon	Honda
	Triumph	Lot	Lotus	Tal	Talbot (formerly	Jaw	Jawa
BMW	BMW	LR	Land-Rover		Chrysler)	Kaw	Kawasaki
Bri	Bristol	Mas	Maserati	Toy	Toyota	Lam	Lambretta
Bui	Buick	Maz	Mazda	Vau	Vauxhall/	Mal	Malaguti
Che	Chevrolet	MB	Mercedes Benz		Bedford	Mbc	Moto Becane
Cit	Citroen	MG	MG	Vlo	Volvo/Daf	Mgz	Moto Guzzi
Col	Colt	Mgn	Morgan	VW	Volkswagen	Mmr	Moto Morini
Dai	Daihatsu	Opl	Opel			MZ	MZ
Dat	Datsun	Peu	Peugeot			Puch	Puch
DJ	Daimler/Jaguar	Por	Porsche	**Motorcycles**		Suz	Suzuki
Fer	Ferrari	Rar	Range Rover	Bat	Batavas	Tri	Triumph
Fia	Fiat	Rel	Reliant	Bet	Beta	Ves	Vespa
		Ren	Renault	BMW	BMW	Yam	Yamaha

As we have already said, driving on a motorway is different from any other road: different techniques must be employed, different regulations apply and different signs must be understood.

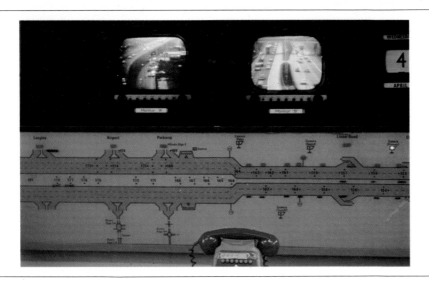

Some motorways are monitored by closed circuit television

The usual layout of a stretch of motorway is:
Pair of hard shoulders, 10ft wide
Pair of three lane carriageways, 36ft wide
Central reservation, 13ft wide, with safety fence
Bridge clearance, 16' 6" minimum
Marker posts on both sides at 110 yard intervals
Telephone at one mile intervals opposite each other

Regulations
Motorway regulations apply from the start of motorway to the 'end of motorway' sign. The following are not allowed to use motorways:
pedestrians
animals
learner drivers (except HGV learners)
pedal cycles
motorcycles of less than 50cc
invalid carriages
agricultural vehicles
slow moving vehicles (except by special permission)

No parking on the central reservation
No parking or reversing on the carriageways or the hard shoulder.
Parking on the hard shoulder only in an emergency.
No walking on the carriageways or

central reservation except in an emergency.
Vehicles over 3 tons unladen weight, and vehicles towing trailers (including cars towing trailers) are not permitted to use the right hand lane of a 3 lane motorway, except in an emergency.

Direction signs
Directional signs and route confirmatory signs are rectangular, with white lettering on a blue background.

Warning signs
Light signals are used to provide motorway warning signs. These are remotely-operated from the motorway control room and are often switched on in response to a radio call from a motorway police patrol. Motorway signals warn of dangers ahead so it is essential to comply with them and reduce speed even if the cause of the restriction is not obvious. When the danger is past the all-clear sign is flashed up on the light signal.

The operation of these signals is monitored by a computer for safety reasons. This prevents a particular light signal from being switched on without the correct sequence of signals leading up to it also being switched on. For example, if a low

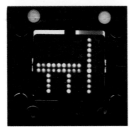

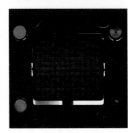

A selection of typical light signals encountered on motorways

speed limit is to be imposed, the computer will ensure that this is preceded by one or more further signals, reducing the speed more gradually.

The computer also enables a continuous record to be kept of which signals were in operation, with details of the time and date. This is designed to prevent a signal from being left on inadvertently.

The signals will be found either on pedestal signs situated on the central reservation or located on an overhead gantry. Warning signs are accompanied by amber lights flashing top and bottom. These lights do not flash for the all-clear sign.

Pedestal signs apply to all lanes of the motorway and contain speed restriction figures and details of lane closures. Gantry signs across the motorway contain, in addition, arrows to warn of the need to change lanes ahead. These signs apply only to the lane immediately below them.

Some of these signals have red stop lights incorporated in them. These flash from side to side, and it is essential to stop when they are switched on. When these stop lights are located on overhead gantries, they apply only to the lane directly below them. When red lights flash on the slip road do not join the motorway.

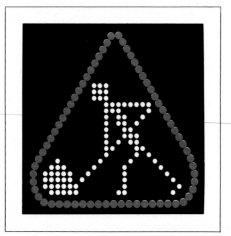

The earliest type of signal used was a pair of amber lights, one above the other, mounted to the left of the hard shoulder. When these flash alternately, they give warning of a hazard such as fog. Motorists are advised to slow down to 30 mph at such a signal.

Message signs
Several motorways have message signs located to the left of the hard shoulder. When not in use, these are blank, but they can be switched on to give warnings of fog, slippery road surface or an accident ahead.

Experimental signs
Some motorways have experimental light signals which are designed to convey some of the variety of information contained on the normal road warning signs. This is not a nationwide system at present but is part of a research programme to help give the motorist more information about why speed restrictions are imposed in the hope that this will lead more motorists to obey them. A limited range of different signs has been developed for this research.

Marker posts
These posts are placed at intervals of 110 yards on the outer edge of each hard shoulder. The sides which face the motorway bear a telephone symbol and an arrow pointing in the direction of the nearest telephone. Each post has a number accompanied by a letter A or B which indicates the carriageway concerned. No telephone is more than ½ mile from any marker post.

Breakdown
In the unfortunate event of a breakdown, it is permissible to use the hard shoulder. When stationary, the hazard warning lights should be switched on, and a warning triangle placed at least 100 yards behind the car.

Great care should be taken when getting out of the car. The driver's door should not be used as it is so close to the carriageway of fast moving traffic. It is preferable to get out of the car through the nearside door. Make sure that passengers, especially children, remain in the car.

After walking to the nearest telephone, a phone call can be made to the motorway control room. The police will take details of the telephone number, the registration number, make, model, year and colour of your car, information about the breakdown, your name and address, and your membership number of a motoring organisation. They will then either ask you to ring them back, or they will ring you back to tell you what action they have taken.

If you are a member of a motoring organisation, the police pass details of the breakdown direct to them. If not, it will be necessary to pay the garage sent out by the police.

It is unwise to leave your car on the hard shoulder for more than a few hours. It is a very hazardous position, and the police move vehicles left in such a place as quickly as possible, charging the owners for doing so. They will also remove your car if you have been involved in a collision or are blocking a carriageway. When rejoining the motorway use the hard shoulder to build up speed before merging into a suitable gap in the traffic.

Motorway junctions are designed to ease the flow of traffic onto and off the motorway. As the general traffic speed is usually quite fast, the slip roads are long enough to allow joining and leaving vehicles plenty of room to build up or lose speed.

Joining the motorway

The entry onto a motorway is made along a two-lane slip road which leads into an acceleration lane. Some slip roads have tight corners, so do not speed up too quickly. As you drive down the slip road watch the traffic in the nearside lane of the motorway, looking for gaps. Long before you reach the point of merging with it, check the mirrors and switch on the right indicators. Look over your right shoulder for gaps in the traffic and match your speed to that of the other vehicles. When a suitable gap occurs, move smoothly into the nearside lane, and cancel the right indicators.

Should traffic on the motorway be so dense that you reach the end of the slip road before a suitable gap has presented itself you must be prepared to stop. You should not try to force your way into the traffic stream.

Once safely on the motorway, accustom yourself to motorway speeds before building up to your desired cruising speed.

When overtaking watch carefully for traffic approaching from behind. It may well be going fast.

Leaving the motorway

On the approach to a junction where you intend to turn off, check the mirrors, switch on the left indicators and move into the nearside lane as early as possible, between the half-mile signpost and the 300 yard countdown sign.

The adjustment from motorway speeds to the halt likely to be required at the junction roundabout needs to take place over a relatively short distance. It is better to glance at the speedometer once or twice on the exit slip road than to rely on one's subjective assessment of speed. Even after joining the subsequent non-motorway road, check your speed.

Telephones

Motorway telephones are situated at regular intervals along the motorway, usually one mile apart. They are placed directly opposite one another to discourage motorists from walking across the carriageways to reach them – an act which is illegal as well as being highly dangerous. The marker posts beside the hard shoulder carry an indication of the direction in which the nearest telephone may be found.

These telephones are for use in emergencies only. They are connected to the motorway control room, and no other calls can be made from them. They are not locked and no money is needed.

Several different styles of telephone are in use, but they all follow the same general principle. The old style telephones work only in one direction to the control room, but with the newer types, it is also possible for the control room to ring back to one of the telephones. They can then tell you what action they have taken.

Direction signs

The approach to every motorway is marked by the international symbol for a motorway, together with its number. From this point until the end of the motorway all motorway regulations (see page 9) apply. A similar sign crossed with a red bar indicates the end of the motorway and its regulations.

Some approach roads to motorways have flashing light signals. If the lights flash they will be either amber, flashing from top to bottom, and will indicate lane closures or an advised maximum speed. Alternatively they may be red lights, flashing from side to side and in that case do not pass the signal. You should not join the motorway at that point but should seek another route to a different motorway junction.

Once you have passed the sign at the start of the motorway you cannot stop or turn round. If you have made a mistake you must continue on the motorway to the next junction and then return to your original route. It is also illegal to stop for hitch-hikers on the motorway.

One mile from each junction the direction sign shows the junction number and the most important road leading away from that junction.

The sign at half a mile from the junction repeats the junction number and gives more information including some of the main destinations reached. At this point, if you are planning to leave the motorway you should move into the left hand lane, after checking mirrors and signalling. Then you can start to lose speed in preparation for the turn off. The start of the deceleration lane leading to the exit slip road is indicated by countdown markers showing the distance in hundreds of yards to the exit. At the first of these, the 300 yard marker, you should switch on the left indicator (after checking the mirror) and prepare to leave the motorway. Do not move onto the hard shoulder and, on no account, overtake any other vehicle in the left hand lane, however slowly they may be going. Lose speed in the deceleration lane.

Service areas

Signs indicating service areas are placed one mile before the turn-off. In some areas there are long distances between service areas, and it is preferable to stop at the first convenient one, particularly if you have been driving continuously for over two hours. The risks of fatigue are considerable after more than two hours motorway driving. This can lead to drowsiness and a possible accident so it is essential to stop occasionally.

The approach to a service area needs great care. In the car park vehicles and pedestrians will be moving in all directions. You will need to lose almost all your speed on the approach slip road, so that you are not going too fast when you enter the car park. Because it is difficult to adjust quickly from the high motorway speeds to the much lower ones required in a service area use your speedometer to get an idea of your speed, rather than rely on your subjective impressions.

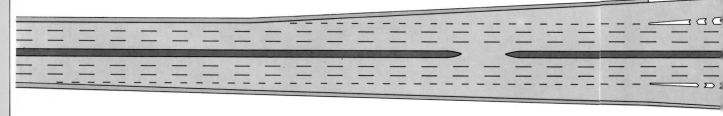

The slip road itself has a confirmatory direction sign, repeating the junction number, the details of the roads and the destinations reached from that junction.

Road markings on the approach to a junction include large arrows indicating the start of the deceleration lane. That lane is bounded by a dashed white line incorporating green reflective markers. The same arrangement will be found on the acceleration lane of the slip road when joining the motorway. The division between the slip road and the motorway is confirmed by an area hatched with diagonal white lines. This indicates that all vehicles should keep off the area.

The motorway junction number shown on the sign is also marked on all modern road maps. Some junctions have the junction number painted on the motorway road surface at the approach to the exit.

Diversion signs

From time to time traffic may have to be diverted off a section of the motorway, to return to it farther along its length. The cause may be roadworks or an accident. A system of signs has been devised in which the diversion route is indicated by a geometric symbol in yellow with a black border. These symbols include

circle, square, triangle and diamond. Each route is given its own symbol.

To keep to a diversion route, all that is necessary is to follow the yellow and black symbols wherever they appear.

Motorways have a better safety record than other roads, and provide fast communication links over long distances. Their freedom from interruptions such as roundabouts, traffic lights and sharp bends means that continuous travel is possible for long distances. Large volumes of traffic can be carried provided everyone observes good lane discipline. Long distances on motorways can prove monotonous, so regular breaks for refreshment and exercise should be taken. Never stop to rest on the hard shoulder but pull off the motorway altogether either onto another road or into a service area.

Car preparation
The strains on a car during a motorway journey can be greater than on other journeys because of the higher continuous speeds involved and the longer distances. A breakdown can be

When preparing your car for a motorway journey be sure to check tyres, petrol, oil and water

an expensive annoyance on a motorway, and reliability is essential.

Lane discipline
This is one of the most crucial aspects of motorway driving. Drivers who travel in the wrong lane or who change lanes without warning and without sufficient care create greater dangers on motorways than on other types of roads. A common but dangerous misconception is that the right hand lane, especially on three lane motorways, is a fast lane. This is not the case, it is an overtaking lane only and is not to be used for continuous

Gantry signs
Junctions in urban areas, or where motorways merge and separate are usually heralded by gantry signs. These contain the junction number and direction information relevant to the lanes below them. They also frequently carry flashing light signals, and are found at distances of $\frac{2}{3}$ mile and $\frac{1}{3}$ mile from the junction.

motoring at 70 mph. Some classes of vehicles are completely prohibited from using the right hand lane of a three lane motorway. These include cars towing caravans or trailers and all goods vehicles with an unladen weight of more than three tons.

The left hand lane should normally be used on two and three lane carriageways. Occasionally it is filled with slow, heavy traffic, for example on some hills.

The centre lane (or right hand lane where there are only two) is used for

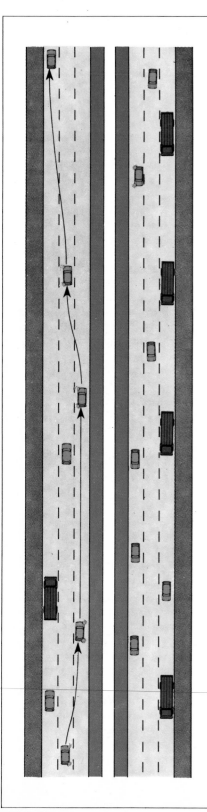

overtaking the traffic in the left hand lane. Where there is a stream of slower traffic in the left hand lane, however, it may be preferable to remain in the centre lane until all the vehicles have been passed, rather than weaving in and out of the lanes. As soon as possible, however, return to the left hand lane. Do not stay in the centre lane if the left hand lane is empty.

All lane changes need to be planned well in advance because of the speeds involved. Always look well ahead so that the need to overtake another vehicle is realised in good time. Carefully check in your mirrors for any traffic approaching from behind. Keep checking this during the overtaking manoeuvre as it is easy to underestimate the speed of following traffic.

Braking distances

A consequence of the fact that speeds are higher on motorways is that braking distances will be greater. It is not always realised just how great these distances become. At 70 mph on a dry road with tyres, car, and driver in first class condition, it takes over 300 feet to stop completely. This is equivalent to about 23 car lengths.

If the roads are wet, the tyres are worn, or the driver's reactions are slow, this distance becomes more than 600 feet, or 46 car lengths. These very long braking distances serve to emphasise the importance of staying alert and keeping adequate spacing between vehicles on motorways.

Consideration for others

An important part of safe driving on motorways is careful consideration for the other road users there. Accurate anticipation of other drivers' moves is essential.

Always try not to baulk other vehicles when driving on a motorway. Before beginning any overtaking manoeuvre, consider whether your car's acceleration is fast enough to take you past the vehicle in front and back into the nearside lane without holding up other, faster traffic wishing to overtake you.

When approaching a motorway junction it is frequently possible to see traffic travelling along the approach slip road intending to join the motorway. It makes their entry onto the motorway simpler if you move from the left hand lane to the centre lane (having checked the mirrors and signalled correctly) well before reaching them.

Lorry drivers are grateful for consideration on motorways. Heavy goods vehicles are forbidden to use the right hand land on a three lane motorway, and must make the best progress they can on the other two lanes. These heavy vehicles sometimes have a separate 'crawler' lane reserved for them on hills. Cars towing caravans are also banned from using the right hand lane on three lane motorways.

Night driving

Dipped headlights should generally be used at night and in adverse weather conditions when travelling on motorways. In spite of the large distance between carriageways it is still possible to dazzle oncoming motorists if main beam lights are used. Always dip your lights if the possibility of dazzle exists either to oncoming drivers or to vehicles ahead of you by reflection in their mirrors.

The system of reflective studs used on motorways is red for the left hand

side, with green at junctions, white along lane markings and amber on the right hand side.

Fatigue

Driving for long distances on motorways can be monotonous, and it can lead to drowsiness and fatigue. Try to prevent this from happening by having adequate rest before starting a journey. Do not start a long journey when you are tired at the end of a day and also avoid driving in the early hours of the morning, since general body alertness is at its lowest point then. Arrange your seat as comfortably as possible. Keep the car well-ventilated, listening to the radio occasionally, checking instruments regularly, and constantly shifting your gaze.

Avoiding heavy meals before or during a journey will lessen the chances of fatigue setting in, as will taking regular breaks, at least every two hours, and more often if you feel like it.

Should you feel yourself becoming drowsy, you should realise that you cannot stop on the hard shoulder and sleep. Get off the motorway at the next exit, however, or at the first service area. Park the car in a safe place off the road, and either sleep there, or take a few minutes' break with some brisk exercise to get the blood circulating again.

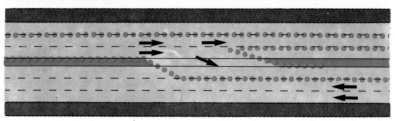

Contra-flow systems are now well-known – select the correct lane early and stay in it.

Roadworks

Most motorways are subject at one time or another to maintenance or improvement works. These can sometimes be substantial undertakings lasting for years. Reduce speed gradually on the approach to road works so that there is no need for last-minute braking.

All roadworks are preceded by sets of warning signs and, sometimes, flashing light signals. In nearly every case high speeds are impossible past roadworks, so you should slow down as soon as the first indication of roadworks is displayed.

When the roadworks are extensive, some lanes of the motorway may be switched onto the other carriageway. This sometimes means you have to make an early decision about the junction at which you need to leave the motorway, since that particular stream may be separated some miles before the junction. This emphasises the importance of planning your route so that you know which junction number you need to take in order to leave the motorway.

Early warning of motorway road works leading to delays is usually published in the press and broadcast on the radio.

Tyres	Stopping distance in feet	100	200	300	400	500	600	700	800	900	1000	1100	1200	1300	Type of road
															Thinking distance 60 feet
Good tread pattern			190 feet												Dry smooth concrete
Good tread pattern				310 feet											Just wet concrete or polished asphalt
Tread depth 1 mm						460 feet									Just wet concrete or polished asphalt
Tyre grooves filled with oil/ rubber sludge									760 feet						Just wet concrete or polished asphalt
Tread depth 2 mm					360 feet										1 mm water on road surface
Near bald														1,360 feet	1 mm water on road surface
Crossply tyre Tread depth 4 mm						490 feet									2 mm water on road surface
Radial ply tyre Tread depth 4 mm					420 feet										2 mm water on road surface
Crossply tyre Tread depth 4 mm											990 feet				4 mm water on road surface
Radial ply tyre Tread depth 4 mm						570 feet									4 mm water on road surface

Weather hazards on the motorway

Adverse weather conditions can make driving more difficult and much less pleasant on any road. On motorways the problems are usually magnified – the road is wider, and therefore more often exposed, and the concentration of traffic is greater.

Before embarking on a long journey, a little forward planning is always advisable, particularly with regard to the weather conditions along the route. National weather forecasts on the radio and television (including Ceefax and Oracle) will give a rough idea of what to expect, but more detailed regional information can be obtained by telephoning any of the numbers listed on page 21. During the journey up-to-date information can be obtained by tuning in to the local radio stations along the way – details of the stations and their wavelengths can be found on pages 22 and 23.

Of course, bad weather will not only affect the motorways – any alternative route in a given area will also suffer, but the weather could present a particular hazard to motorway users, such as high cross winds, which drivers would prefer to avoid. Here we give details of such trouble spots on the motorway network so that steps can be taken to avoid them whenever necessary. We don't claim that these are the only places where difficulties might arise – they are just the most likely. Now and then certain stretches of motorway are judged so hazardous because of the weather that they are closed altogether. In these cases, police signs will indicate an alternative route, but it is generally the case that a backlog of traffic will already have built up. Foreward planning and the anticipation of such trouble spots will save this kind of frustrating delay to a journey.

Snow and ice

If wintry conditions are widespread over the intended journey and are likely to remain so, by far the best policy is to postpone your journey. The popular phrase of the weather bulletins is 'Don't travel unless it is absolutely necessary'. If you do decide to go, make sure the car is in tip-top condition and always set off prepared for the worst. Then travel hopefully and carefully, with the consolation that motorways and main roads are the first to receive the attention of the local authorities' snow squads.

If snow showers and icy conditions are isolated, these are the trouble-spots to watch out for:

M1	Junctions 22–23
M4	Junction 14
M6	Junctions 37–43

M8	Junctions 4–6 and 27–28
M62	Junctions 21–22
M74	Junctions 2–3
M90	Junctions 3–6
A1(M)	Darlington–Durham

How to cope

Before setting off put a few extras into the car in case you should get stuck: a warm coat and boots, a hot drink in a thermos flask, a shovel and some sacking (to put under the wheels to assist grip), damp-start and de-icer, a torch and a tow-rope. Also make sure the lamps are all clean and that there is anti-freeze in the windscreen washers.

When driving on icy or snow-covered roads, the golden rule is to keep calm and use all the controls gently, especially the brakes. Try to anticipate icy stretches of road and other hazards before you reach them and get your speed and your gears right in plenty of time. Keep your distance from other drivers in case you have to react to an emergency. Black ice is difficult to detect, but you can tell if it is present on the road if there is no 'swishing' sound from the tyres and if the steering feels unnaturally light. If your car does skid, don't panic and

don't jam your foot on the brakes, as this will lock the wheels and you will lose control. Take your foot off the accelerator and straighten up the steering. Never brake and steer at the same time. To reduce speed on a slippery surface, brake gently, release the brakes and steer in the right direction, then brake gently again.

It is not a good idea to change down to a lower gear to reduce speed because the action of letting in the clutch has a braking effect on the engine which is transmitted to the driving wheels. Uphill or downhill, remember that you will probably need a low gear. Engage it before you embark on the hill so that you can keep to a constant, safe speed while ascending or descending. If there is other traffic already on the hill, it may be best to wait until you can see that you have a clear run. Remember that going downhill may be even more tricky than going uphill.

Fog

Fog is another great deterrent to drivers and a very difficult condition to adjust to. Drivers are often accused of travelling too fast, but it is less likely to be recklessness and more the effect of the fog hampering one's sense of speed. The fact that car lights and street lighting are reflected back adds to the disorientation and robs drivers of the familiar landmarks by which they assess their speed and direction.

Fortunately, on motorways one of the worst hazards of driving in fog is eliminated – negotiating cross-roads and right turns across invisible oncoming traffic. The main problem is from disorientation from one's surroundings and the speed of other motorists, particularly those behind. Fog forms from moisture in the air coupled with a drop in temperature, so obvious danger areas are near rivers, lakes, gravel pits, meadows and some woodland. The map (right) shows the areas of the country worst affected by fog and danger areas on the motorways are:

M1	Junctions 7–12 and 22–28, particularly around junction 26
M3	Junctions 3–7
M4	Junctions 10–12, 14–17 and 19–21
M5/50	Junctions 3–4, 8 and four miles north of junction 22.
M6	Junctions 1–4, 10–11, 14–15 and the area north of junction 15.
M8	Junctions 2–3, 4 and 29–30
M9	Junctions 2–3
M62	Junctions 21–22
M74	Junctions 4–5
M90	Junctions 2–4 and around the Forth Road Bridge
A1(M)	Around Newcastle upon Tyne

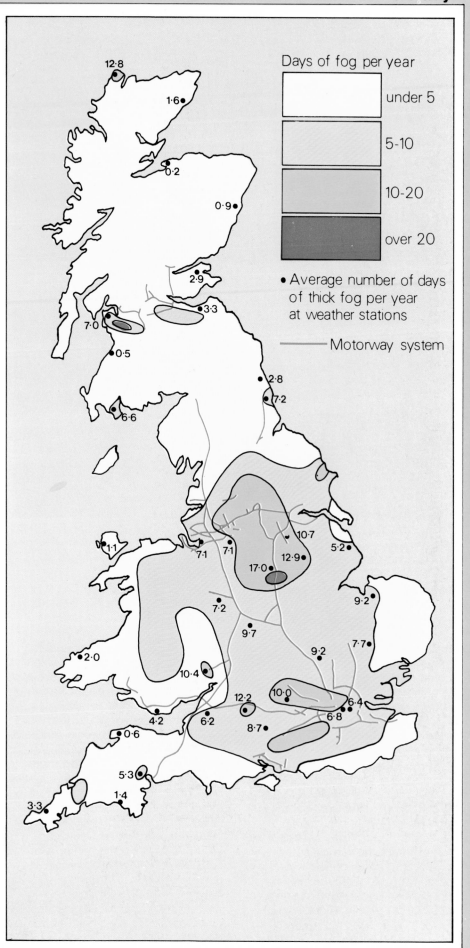

Days of fog per year

under 5

5-10

10-20

over 20

• Average number of days of thick fog per year at weather stations

—— Motorway system

Dipped headlights and a slow speed are essential in fog. Always conform to speed restrictions.

How to cope

Fog is the hazard that drivers fear the most – and with reason. Not only does it strain all the senses, but it produces misleading illusions – distorting impressions of speed, distance and colour; familiar landmarks disappear, intensifying a feeling of disorientation. Care should be taken at traffic lights; fog can cause red to appear amber, amber white and green to disappear altogether.

The use of dipped headlights in fog is mandatory – both to let other road-users see your vehicle and to improve your own range of vision. Headlights on full beam are not much help because the angle and spread of the beam is such that the fog particles will reflect back into the driver's eyes and dazzle him. Sidelights on their own are not sufficient: your vehicle cannot be seen by other drivers soon enough. Remember to keep the lenses clean, and if you have to stop to clean them,

find a safe place to do so. Fog lights help, but if you have only one fitted, always use it in conjunction with dipped headlights to avoid being mistaken for a motorcycle. Rear fog lights are also useful, but they should not be wired into the brake-light circuit, and they should give about the same intensity of light as the brake lights. Keep your speed down, so that you are sure you can stop within your field of vision, even if it means driving at no more than 5mph. Do not be tempted to follow another vehicle's tail lights – the other driver may be driving too fast for safety. Observe motorway speed restrictions, and if you have to overtake, do so with extreme care: if you have been driving behind a lorry, for example, you may have gained a false impression of the thickness of the fog ahead. Open a window, both to prevent the car from misting up inside and to help you hear other traffic. If you do have to stop, always try to get off

the carriageway. If this is impossible, switch on the hazard warning lights and put out a warning triangle 150 yards behind the car.

Cross-winds

In creating the motorway system, wide paths have been cut across some fairly exposed country and it is on these roads that the problem of high cross-winds is the greatest. Instances of cross-winds require from the driver a greater degree of concentration and quicker responses. Sudden gusts of wind can cause swerving and in a strong but steady wind, a similar effect can be experienced in a sudden lull. Overtaking large, high-sided vehicles can be particularly hazardous and cars towing caravans will be affected more than most by high winds. The addition of a spoiler can be a helpful aid to stability. The following motorway sections are particularly susceptible to high cross-winds:

M1	Junctions 29–30
M2	Junctions 2–3 (Medway Viaduct) and 5–7
M3	Junctions 3–7
M4	Junctions 18–19 and the Severn Bridge
M5	Junctions 14–15
M6	Junctions 2–4, particularly on the Longford Viaduct (Junction 3), 12–13 and 37–43
M8	Junctions 3–5 and 27–30
M9	Junctions 2–3
M90	Forth Road Bridge and all exposed areas particularly junction 3
M62	Junctions 21–22
M74	Junctions 1–2 and 4–5
A1(M)	Darlington–Durham

How to cope

Although windy conditions may not at first appear to be a hazard, statistics do show that on sections of motorways which are exposed to high winds, more accidents occur on days when wind conditions are worse than average. One of the difficulties is that a driver on a long journey may not be aware that high winds have sprung up since he started out. Observation of the terrain ahead and of the behaviour of cars in front of you may give you forewarning of the presence of high

vehicles needs care too. As you come alongside such a vehicle you are, momentarily, sheltered and because you have been steering a course to compensate for the effect of side wind, you may at this point feel that you are being sucked in towards the other vehicle. When you leave its shelter, however, the effects of wind will be felt again. Roof racks affect the car's handling in windy weather. Fasten luggage securely and as low as possible to optimise air-flow over the car. Towing a caravan requires

special care in high winds. If the caravan is not properly loaded, strong cross-winds can cause the phenomenon known as 'snaking', which begins as a gentle weaving motion, but quickly builds up until car and caravan are swaying out of unison and become difficult to handle. Do not steer against the snaking, or brake sharply, but lose speed slowly and brake gently until control is restored. If snaking happens at low speed, accelerate briefly and gently until the vehicles straighten out again.

winds. If the vehicles ahead suddenly veer to one side, it is likely that they are experiencing strong cross-winds. If you can see that you are about to enter an exposed stretch of road, you may expect cross-winds on a bad day. The parapets of motorway bridges can also deflect wind, producing gusts that seem to come from all directions. Although drivers must respond quickly, reactions should not be so violent as to cause a skid. Be careful not to over-correct your steering.

Rain
Rain is something that we are quite used to in this country and it doesn't normally present too many driving difficulties. However, visibility can be seriously affected on motorways, particularly when following heavy lorries. The amount of spray

generated by their multiple wheels can be a real hazard and certainly makes driving extremely unpleasant. Keeping a safe distance from the vehicle in front is the obvious answer but this does not solve the problem of spray from traffic overtaking you.

Modern tyres can handle quite strong cross-winds and it is more effective to work with the tyres than against them. You should, generally, reduce speed in high winds. Passing high-sided

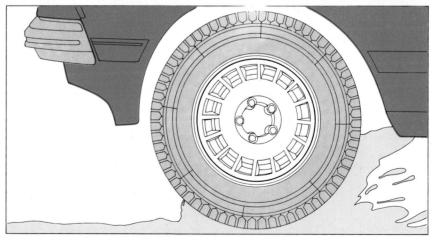

Aquaplaning (shown above, diagrammatically) is caused by the build up of a cushion of water in front of the tyres, when driving fast on surface water.

Heavy rain can overload wiper blades, allowing an almost continuous sheet of water to flow over the screen. Rain on mirrors can also distort your view of traffic behind.

How to cope

Rain means reduced visibility, a deterioration in the car's stopping ability and increased risks of skidding and aquaplaning. Rain is most dangerous when it falls after a long period of dry weather, when the water meets oil and rubber deposits on the road surface and produces a mixture almost as slippery as ice. If the soles of your shoes are wet, they may slip on the control pedals, so remember to scuff them on carpets or matting to give you a safer grip. The basic rules are: reduce speed to allow you to manoeuvre safely on bends or in mixed traffic; switch on dipped headlights so that you can easily be seen; use windscreen washers and wipers, demisters and screen heaters; do not brake or swerve violently in case you provoke a skid. Aquaplaning is the most disconcerting hazard, and it is when you encounter this phenomenon that the importance of good tyres with an adequate depth of tread becomes apparent. Aquaplaning occurs when the vehicle is driven at speed into surface water and a cushion of water builds up ahead of the tyre. If the tread grooves do not allow sufficient water to pass

Bright sunlight

Not normally considered 'bad weather', a bright sun can still be a hazard if you are driving directly into it. Eye strain can be considerable over a long journey and visibility impaired. The worst culprit is reflected sunlight which has become partly or wholly polarised – the eye is extremely sensitive to this polarised light. The winter sun, low in the sky, can be particularly difficult to avoid.

How to cope

A good pair of sunglasses is the answer with lenses that will absorb about 75% of infra-red radiation without impairing general visibility. Make sure that there are no flaws in the lenses that could distort the light passing through. Driving from bright sunlight into a tunnel will cause temporary blindness, best avoided by removing sunglasses before entering the tunnel and using dipped headlights until you emerge.

through them, contact with the road surface is lost and neither brakes nor steering can control the vehicle. Take your foot off the accelerator to lose speed. If there is $\frac{1}{8}$ inch (3mm) of surface water on the road aquaplaning may occur at 45mph if your tyres have 1mm of tread depth (the legal minimum). Lowering the tyre pressure, far from improving matters, simply reduces the speed at which aquaplaning may occur. Surface water tends to accumulate in the slow lanes of motorways in the tracks left by heavy vehicles. On motorways, remember that visibility in rain is further reduced by the spray thrown up from the tyres of heavy vehicles and great care should be exercised when overtaking them.

REMEMBER

Use your seat belts as required by law – and make sure your front-seat passenger uses one too.

If you have young children, have child-safety seats fitted and child-proof locks on the doors. Do not allow young children or pets to roam about the interior of the vehicle.

Never drink and drive.

A heavy meal will tend to make you feel drowsy. Several light snacks and short periods of exercise are better than one big meal.

Break your journey and stretch your legs at least every two hours.

Never start on a long journey at the end of the day.when you are feeling tired.

If you feel drowsy, open a window, put the radio on or talk to your passengers until you can turn off the motorway and find a safe place, off the road, for a rest. Do not use the hard shoulder except in emergency.

A warm fug inside the vehicle induces drowsiness, so keep the vehicle well ventilated, even on cold days.

If you have to take drugs – even ones freely available such as aspirin – check with your doctor for any possible effect on your driving.

If you feel ill, get off the road as soon as possible.

Keep your vehicle in good condition: clean windows and lenses often and keep the washer bottle topped up; check tyre pressures and tread-depth regularly; make sure your spare tyre is serviceable; carry a set of spare bulbs, fuses etc. with you; check the oil gauge, the fan belt, the water and the battery fluid levels before setting out on a long journey; always have your vehicle serviced at the proper intervals.

Motoring information by phone

Recorded reports on road conditions are included in the British Telecom 'Traveline' service covering nine separate areas of the United Kingdom. The reports are available by telephone at all times and give information about major roads affected by adverse weather, road works, special events, etc.

The telephone numbers are listed below, with area dialling codes shown in brackets. **A national summary of motorway conditions is available from London: telephone 01-246-8031.**

70 miles of London

LONDON	01-246 8021
BEDFORD	(0234) 8021
BISHOPS STORTFORD	(0279) 8021
BRIGHTON	(0273) 8021
CHELMSFORD	(0245) 8021
COLCHESTER	(0206) 8021
GUILDFORD	(0483) 8021
HIGH WYCOMBE	(0494) 8021
LUTON	(0582) 8021
MEDWAY	(0634) 8021
OXFORD	(0865) 8021
PORTSMOUTH	(0705) 8021
READING	(0734) 8021
SOUTHEND	(0702) 8021
TUNBRIDGE WELLS	(0892) 8021

50 miles of Birmingham

BIRMINGHAM	021-246 8021
COVENTRY	(0203) 8021
HEREFORD	(0432) 8021

70 miles of Bristol

BRISTOL	(0272) 8021
BOURNEMOUTH	(0202) 8021
CHELTENHAM	(0242) 8021
GLOUCESTER	(0452) 8021
SOUTHAMPTON	(0703) 8021
SWINDON	(0793) 8021

50 miles of Liverpool

LIVERPOOL	051-246 8021
MANCHESTER	061-246 8021
BLACKBURN	(0254) 8021
BLACKPOOL	(0253) 8021

Northern England

NEWCASTLE UPON TYNE	(0632) 8021
MIDDLESBROUGH	(0642) 8021

South & West Yorkshire

LEEDS	(0532) 8021
BRADFORD	(0274) 8021
DONCASTER	(0302) 8021
HUDDERSFIELD	(0484) 8021
SHEFFIELD	(0742) 8021

South Wales

CARDIFF	(0222) 8021
NEWPORT (GWENT)	(0633) 8021

Scotland

EDINBURGH	031-246 8021
GLASGOW	041-246 8021

Northern Ireland

BELFAST	(0232) 8021
BANGOR	(0247) 8021

Meteorological office weather centres

Members of the public may telephone or call at the centres listed below for information about local, national and Continental weather forecasts. Other meteorological office telephone numbers are given in the prefaces to telephone directories. **NB Meteorological offices do not give information about road conditions.**

London
284–286 High Holborn 01-836 4311

Glasgow
118 Waterloo Street 041-248 3451

Manchester
56 Royal Exchange 041-832 6701

Newcastle
7th Floor, Newgate House, Newgate Street (0632) 326453

Nottingham
Main Road, Watnall (0602) 384091

Southampton
160 High Street, Below Bar (0703) 28844

BBC Local Radio

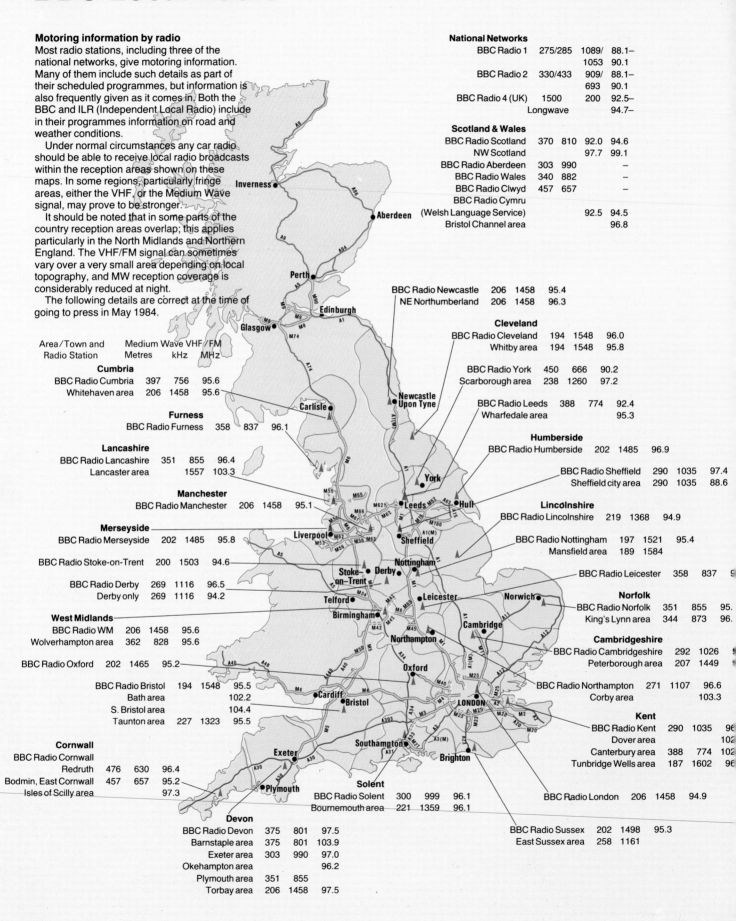

Motoring information by radio

Most radio stations, including three of the national networks, give motoring information. Many of them include such details as part of their scheduled programmes, but information is also frequently given as it comes in. Both the BBC and ILR (Independent Local Radio) include in their programmes information on road and weather conditions.

Under normal circumstances any car radio should be able to receive local radio broadcasts within the reception areas shown on these maps. In some regions, particularly fringe areas, either the VHF, or the Medium Wave signal, may prove to be stronger.

It should be noted that in some parts of the country reception areas overlap; this applies particularly in the North Midlands and Northern England. The VHF/FM signal can sometimes vary over a very small area depending on local topography, and MW reception coverage is considerably reduced at night.

The following details are correct at the time of going to press in May 1984.

Area/Town and Radio Station	Medium Wave Metres	kHz	VHF/FM MHz
Cumbria			
BBC Radio Cumbria	397	756	95.6
Whitehaven area	206	1458	95.6
Furness			
BBC Radio Furness	358	837	96.1
Lancashire			
BBC Radio Lancashire	351	855	96.4
Lancaster area		1557	103.3
Manchester			
BBC Radio Manchester	206	1458	95.1
Merseyside			
BBC Radio Merseyside	202	1485	95.8
BBC Radio Stoke-on-Trent	200	1503	94.6
BBC Radio Derby	269	1116	96.5
Derby only	269	1116	94.2
West Midlands			
BBC Radio WM	206	1458	95.6
Wolverhampton area	362	828	95.6
BBC Radio Oxford	202	1465	95.2
BBC Radio Bristol	194	1548	95.5
Bath area			102.2
S. Bristol area			104.4
Taunton area	227	1323	95.5
Cornwall			
BBC Radio Cornwall			
Redruth	476	630	96.4
Bodmin, East Cornwall	457	657	95.2
Isles of Scilly area			97.3
Devon			
BBC Radio Devon	375	801	97.5
Barnstaple area	375	801	103.9
Exeter area	303	990	97.0
Okehampton area			96.2
Plymouth area	351	855	
Torbay area	206	1458	97.5
Solent			
BBC Radio Solent	300	999	96.1
Bournemouth area	221	1359	96.1

National Networks

BBC Radio 1	275/285	1089/ 1053	88.1– 90.1
BBC Radio 2	330/433	909/ 693	88.1– 90.1
BBC Radio 4 (UK)	1500 Longwave	200	92.5– 94.7–

Scotland & Wales

BBC Radio Scotland	370	810	92.0	94.6
NW Scotland			97.7	99.1
BBC Radio Aberdeen	303	990		–
BBC Radio Wales	340	882		
BBC Radio Clwyd	457	657		–
BBC Radio Cymru (Welsh Language Service)			92.5	94.5
Bristol Channel area				96.8

BBC Radio Newcastle	206	1458	95.4
NE Northumberland	206	1458	96.3
Cleveland			
BBC Radio Cleveland	194	1548	96.0
Whitby area	194	1548	95.8
BBC Radio York	450	666	90.2
Scarborough area	238	1260	97.2
BBC Radio Leeds	388	774	92.4
Wharfedale area			95.3
Humberside			
BBC Radio Humberside	202	1485	96.9
BBC Radio Sheffield	290	1035	97.4
Sheffield city area	290	1035	88.6
Lincolnshire			
BBC Radio Lincolnshire	219	1368	94.9
BBC Radio Nottingham	197	1521	95.4
Mansfield area	189	1584	
BBC Radio Leicester	358	837	9
Norfolk			
BBC Radio Norfolk	351	855	95.
King's Lynn area	344	873	96.
Cambridgeshire			
BBC Radio Cambridgeshire	292	1026	9
Peterborough area	207	1449	
BBC Radio Northampton	271	1107	96.6
Corby area			103.3
Kent			
BBC Radio Kent	290	1035	96
Dover area			102
Canterbury area	388	774	102
Tunbridge Wells area	187	1602	96
BBC Radio London	206	1458	94.9
BBC Radio Sussex	202	1498	95.3
East Sussex area	258	1161	

22

IBA Local Radio

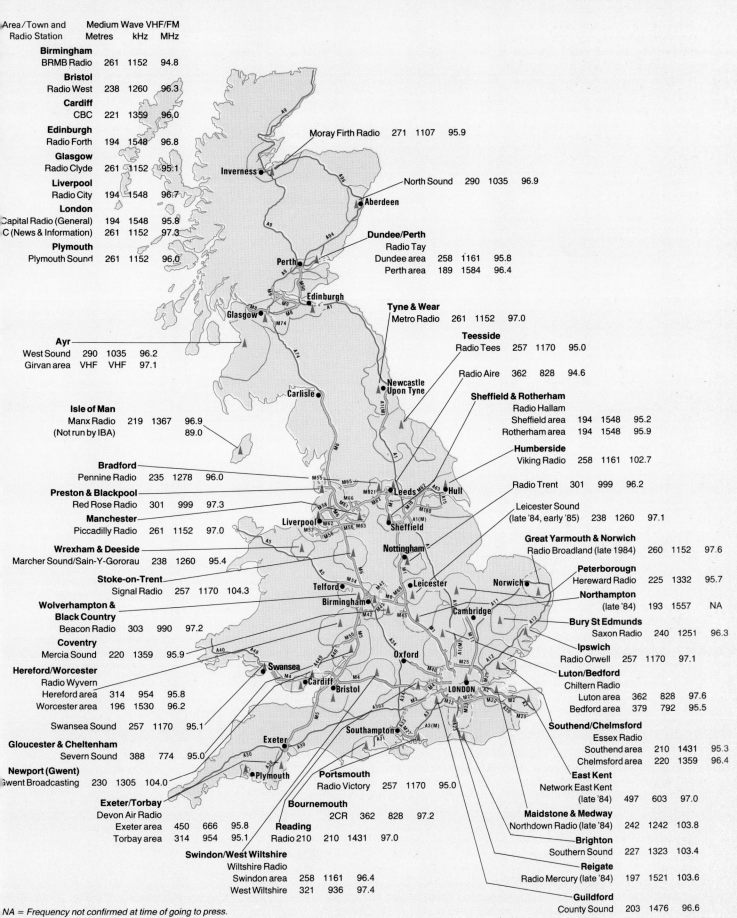

Area/Town and Radio Station	Medium Wave Metres	kHz	VHF/FM MHz
Birmingham			
BRMB Radio	261	1152	94.8
Bristol			
Radio West	238	1260	96.3
Cardiff			
CBC	221	1359	96.0
Edinburgh			
Radio Forth	194	1548	96.8
Glasgow			
Radio Clyde	261	1152	95.1
Liverpool			
Radio City	194	1548	96.7
London			
Capital Radio (General)	194	1548	95.8
LBC (News & Information)	261	1152	97.3
Plymouth			
Plymouth Sound	261	1152	96.0

Moray Firth Radio 271 1107 95.9

North Sound 290 1035 96.9

Dundee/Perth
Radio Tay
Dundee area 258 1161 95.8
Perth area 189 1584 96.4

Tyne & Wear
Metro Radio 261 1152 97.0

Teesside
Radio Tees 257 1170 95.0

Radio Aire 362 828 94.6

Sheffield & Rotherham
Radio Hallam
Sheffield area 194 1548 95.2
Rotherham area 194 1548 95.9

Humberside
Viking Radio 258 1161 102.7

Radio Trent 301 999 96.2

Leicester Sound
(late '84, early '85) 238 1260 97.1

Great Yarmouth & Norwich
Radio Broadland (late 1984) 260 1152 97.6

Peterborough
Hereward Radio 225 1332 95.7

Northampton
(late '84) 193 1557 NA

Bury St Edmunds
Saxon Radio 240 1251 96.3

Ipswich
Radio Orwell 257 1170 97.1

Luton/Bedford
Chiltern Radio
Luton area 362 828 97.6
Bedford area 379 792 95.5

Southend/Chelmsford
Essex Radio
Southend area 210 1431 95.3
Chelmsford area 220 1359 96.4

East Kent
Network East Kent
(late '84) 497 603 97.0

Maidstone & Medway
Northdown Radio (late '84) 242 1242 103.8

Brighton
Southern Sound 227 1323 103.4

Reigate
Radio Mercury (late '84) 197 1521 103.6

Guildford
County Sound 203 1476 96.6

Ayr
West Sound 290 1035 96.2
Girvan area VHF VHF 97.1

Isle of Man
Manx Radio 219 1367 96.9
(Not run by IBA) 89.0

Bradford
Pennine Radio 235 1278 96.0

Preston & Blackpool
Red Rose Radio 301 999 97.3

Manchester
Piccadilly Radio 261 1152 97.0

Wrexham & Deeside
Marcher Sound/Sain-Y-Gororau 238 1260 95.4

Stoke-on-Trent
Signal Radio 257 1170 104.3

Wolverhampton & Black Country
Beacon Radio 303 990 97.2

Coventry
Mercia Sound 220 1359 95.9

Hereford/Worcester
Radio Wyvern
Hereford area 314 954 95.8
Worcester area 196 1530 96.2

Swansea Sound 257 1170 95.1

Gloucester & Cheltenham
Severn Sound 388 774 95.0

Newport (Gwent)
Gwent Broadcasting 230 1305 104.0

Exeter/Torbay
Devon Air Radio
Exeter area 450 666 95.8
Torbay area 314 954 95.1

Swindon/West Wiltshire
Wiltshire Radio
Swindon area 258 1161 96.4
West Wiltshire 321 936 97.4

Portsmouth
Radio Victory 257 1170 95.0

Bournemouth
2CR 362 828 97.2

Reading
Radio 210 210 1431 97.0

NA = Frequency not confirmed at time of going to press.

Motorways under construction

At the time of going to press a number of new developments in the motorway system are still under construction. The completion dates, while not exactly unpredictable, are subject to all kinds of influences which can cause delays – it has even been known for new roads to open ahead of schedule! Below we list the motorway sections which are in progress and give the expected completion date. Up-to-date information can be obtained nearer the time from national and local newspapers, radio (see pages 22–23) and television (including Ceefax and Oracle), or by ringing British Telecom's 'Traveline' service (see page 21).

The M25 takes shape near Leatherhead in Surrey.

M3	Junction 8 (Basingstoke)	Junction 10 (Winchester)	Summer 1985
M4	Junction 33		Winter 1984
M25	Junction 3 (Swanley)	Junction 5 (Sevenoaks)	Spring 1986
M25	Junction 8 (Reigate)	Junction 10 (A3)	Summer 1985
M25	Junction 14 (A3113)	Junction 15 (M4)	Summer 1985
M25	Junction 15 (M4)	Junction 17 (Maple Cross)	Spring 1985
M25	Junction 19 (A405)	Junction 21 (M1 & A405)	Autumn 1986
M25	Junction 21 (A405)	Junction 23 (South Mimms)	Summer 1986
M42	Junction 1 (A34)	Junction with A441 (Alvechurch)	Spring 1985
M42	Junction with A441 (Alvechurch)	Junction with A38 (Bromsgrove)	Early 1986
M42	Junction 4 (M6, Coleshill)	Junction with A453 (Appleby Magna)	Summer 1986
M65	Junction 7 (Clayton-le-Moor)	Junction 6 (Blackburn)	Summer 1985
M74	Junction 1 (Blackwood)	Junction with A70 & A74	Summer 1986

The M25 London Orbital Motorway

The mammoth task of creating what is, in effect, the world's largest bypass is the most important motorway project to be undertaken in this country and completion is likely to be within the next two years or so.

Twenty five years of motorway and trunk road construction throughout Britain had, by the mid 1970s, increased accessibility to the capital and created a huge bottleneck on its already overcrowded roads. The need for a purpose-built orbital route became vital.

Because it encircles such a large area, with a high concentration of habitation all along its route, the planning and approval processes were far from straightforward. Proposed routes changed and changed again, and each section has been the subject of lengthy public enquiry. All the problems and objections had to be resolved before the first earthmover could move in.

The actual construction work has been no less complicated – not merely a case of cutting a swathe through open countryside. Some unusual steps have been taken to set things to rights, such as a special grass/concrete crossing constructed in the Epping Forest section so that the deer can still wander at will; a cricket pitch, destroyed to make way for the Bell Common tunnel, restored; tennis courts and a running track created above Holmesdale tunnel; a thatched cottage, the lodge to Upshire Hall, taken apart and rebuilt to make way for the A121 interchange. A major project of planting such indigenous species as beech, oak, hornbeam and hawthorn, will also help the motorway fit into its surroundings.

Of an eventual 121.5 miles, about 75 miles are complete, part of which now forms an eastern bypass of the capital between the A1(M) and the M20. It is, in fact, now possible to drive from Glasgow to the Kent Channel ports entirely on motorways and dual carriageways (except for a stretch of the A20 to Folkestone). The cost of building the motorway is enormous – projected at £875 million – but it is estimated that by the turn of the century 100,000 vehicles a day will be using the busiest sections, with 200,000 or more passing through the M4/M25 intersection. Access to and between the airports at Heathrow, Gatwick and Stansted will also be greatly improved.

Four service areas are planned, but as yet no firm decision has been made about their location. However, two are likely to be in the vicinity of the Dartford Tunnel and South Mimms, with the others on the western and southern sections.

Legend

Motorway

Motorway under construction

Projected motorway (line fixed)

Primary route

0 10 20 30 40 50 miles
0 20 40 60 80 kilometres

Scale 45 miles to 1 inch

NORTH SEA

CAERNARVON BAY

CARDIGAN BAY

BRISTOL CHANNEL

THE WASH

FIRTH OF FORTH

FIRTH OF CLYDE

SOLWAY FIRTH

25

Route Planning Maps

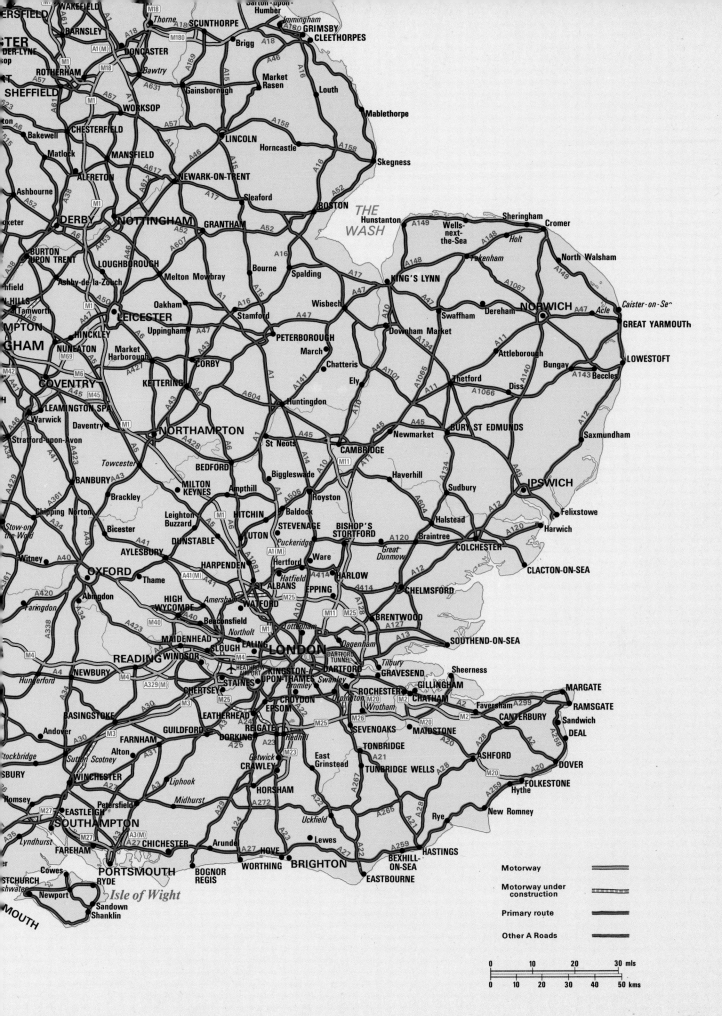

Island
of
Mull

Oban   A85   Tyndrum   Killin

Dalmally   A85   Crianlarich   Lochearnhead   A85   Crieff   PERT

A816   Inveraray   A83   Callander   A9   Auchtera

A85   Dunblane   A91   A971

A815   STIRLING

A82   M9   DUNFER

Lochgilphead   M80   Kincardine-on-Forth

Ardrishaig   Dunoon   Gourock   DUMBARTON   ERSKINE BRIDGE   A80   FALKIRK   A9

GREENOCK   Wemyss Bay   M8   GLASGOW   M73   Linlithgow   M8   Livi

Tarbert   A78   Largs   PAISLEY   AIRDRIE

Islay   A736   M74   MOTHERWELL   A71

HAMILTON

Kintyre   A83   Ardrossan   A78   Strathaven   Lanark

Island   Brodick   Saltcoats   A71   KILMARNOCK   A71   A73   Big

of   Lamlash   IRVINE   A77   A74

Arran   Prestwick   A76   A702

Campbeltown   AYR   Cumnock   A701

Maybole   A713   Sanquhar   A702   Mc

Girvan   A76

A714   A702   New Galloway   DUMFRIES

A77   A712   A75

Newton Stewart   A75

Stranraer   A75   Gatehouse   Castle Douglas
of Fleet

Maryport

WORKINGTON   Co

A5086   Ke

Egremont   A595

Isle of
Man   A3   Ramsey

Peel   Laxey   A2

A4   Onchan

Port   A5   DOUGLAS   BARROW-
Erin   IN-FURNESS

Castletown

FL

BLACK

SO

Holyhead   Anglesey   COLWYN   RHYL   BIRKI
A5   BAY   Prestatyn   BE

Conwy   Abergele   A55

Bangor   A55   A525   Que

Caernarfon   A5   Denbigh

Betws-y-coed   A494   Ruth

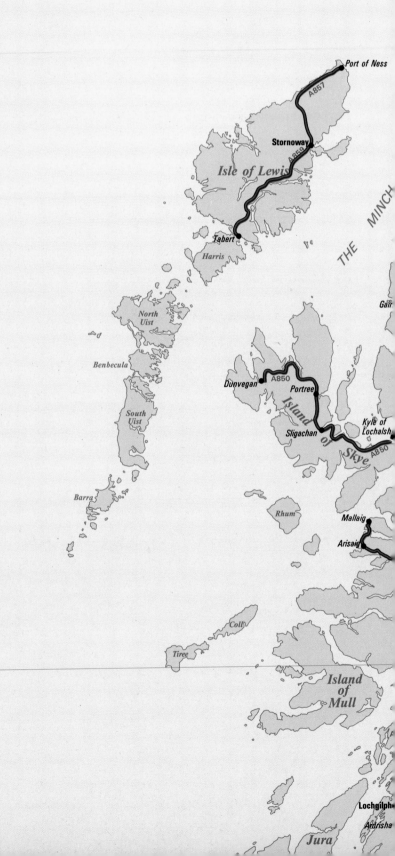

Port of Ness

A857

Stornoway

A850

Isle of Lewis

Tabert

Harris

North
Uist

Benbecula

Dunvegan   A850
            Portree
Island
Sligachan   of   Kyle of
            Skye Lochalsh
                   A850

South
Uist

THE MINCH

Gair

Barra

Rhum

Mallaig

Arisaig

Coll

Tiree

Island
of
Mull

Lochgilph

Ardrisha

Jura

Mileage Chart

The distances between towns on the mileage chart are given to the nearest mile, and are measured along the normal AA recommended routes. It should be noted that AA recommended routes do not necessarily follow the shortest distances between places but are based on the quickest travelling time, making maximum use of motorways or dual-carriageway roads.

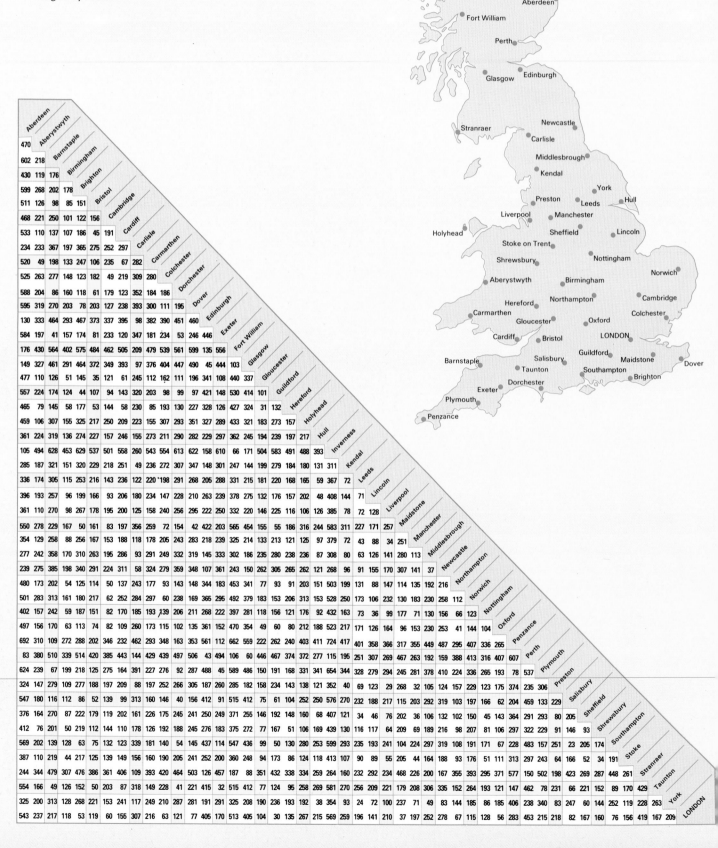

Column order (left to right): Aberdeen, Aberystwyth, Barnstaple, Birmingham, Brighton, Bristol, Cambridge, Cardiff, Carlisle, Carmarthen, Colchester, Dorchester, Dover, Edinburgh, Exeter, Fort William, Glasgow, Gloucester, Guildford, Hereford, Holyhead, Hull, Inverness, Kendal, Leeds, Lincoln, Liverpool, Maidstone, Manchester, Middlesbrough, Newcastle, Northampton, Norwich, Nottingham, Oxford, Penzance, Perth, Plymouth, Preston, Salisbury, Sheffield, Shrewsbury, Southampton, Stoke, Stranraer, Taunton, York.

Place	Distances (in column order above)
Aberystwyth	470
Barnstaple	602 218
Birmingham	430 119 176
Brighton	599 268 202 178
Bristol	511 126 98 85 151
Cambridge	468 221 250 101 122 156
Cardiff	533 110 137 107 186 45 191
Carlisle	234 233 367 197 365 275 252 297
Carmarthen	520 49 198 133 247 106 235 67 282
Colchester	525 263 277 148 123 182 49 219 309 280
Dorchester	588 204 86 160 118 61 179 123 352 184 186
Dover	595 319 270 203 78 203 127 238 393 300 111 195
Edinburgh	130 333 464 293 467 373 337 395 98 382 390 451 460
Exeter	584 197 41 157 174 81 233 120 347 181 234 53 246 446
Fort William	176 430 564 402 575 484 462 505 209 479 539 561 599 135 556
Glasgow	149 327 461 291 464 372 349 393 97 376 404 447 490 45 444 103
Gloucester	477 110 126 51 145 35 121 61 245 112 162 111 196 341 108 440 337
Guildford	557 224 174 124 44 107 94 143 320 203 98 99 97 421 148 530 414 101
Hereford	465 79 145 58 177 53 144 58 230 85 193 130 227 328 126 427 324 31 132
Holyhead	459 106 307 155 325 217 250 209 223 155 320 351 327 289 433 321 183 273 157
Hull	361 224 136 274 227 157 246 155 273 282 329 297 362 245 194 239 197 217
Inverness	105 494 628 453 629 537 501 558 260 543 554 613 622 158 610 66 171 504 583 491 488 393
Kendal	285 187 321 151 320 229 218 251 49 236 272 307 347 148 301 247 144 199 279 184 180 131 311
Leeds	336 174 305 115 253 216 143 236 122 220 198 291 268 205 288 331 215 181 220 168 165 59 367 72
Lincoln	396 193 257 96 199 166 93 206 180 234 147 228 210 263 239 378 275 132 176 157 202 48 408 144 71
Liverpool	361 110 270 98 267 178 195 200 125 158 240 256 295 222 250 332 220 146 225 116 106 126 385 78 72 128
Maidstone	550 278 229 167 50 161 83 197 356 259 72 154 42 422 203 565 454 155 55 186 316 244 583 311 227 171 257
Manchester	354 129 258 88 256 167 153 188 118 178 205 243 283 218 239 325 214 133 213 121 125 97 379 72 43 88 34 251
Middlesbrough	277 242 358 170 310 263 195 286 93 291 249 332 319 145 333 302 186 235 280 238 236 87 308 80 63 126 141 280 113
Newcastle	239 275 385 198 340 291 224 311 58 324 279 348 107 361 243 150 262 305 265 262 121 268 96 91 155 170 307 141 37
Northampton	480 173 202 54 125 114 50 137 243 177 93 143 148 344 183 453 341 77 93 91 203 151 503 199 131 88 147 114 135 192 216
Norwich	501 283 313 161 180 217 62 252 284 297 60 238 169 365 295 492 379 183 153 206 313 153 528 250 173 106 232 130 183 230 258 112
Nottingham	402 157 242 59 187 151 82 170 185 193 139 206 211 268 222 397 281 118 156 121 176 92 432 163 73 36 99 177 71 130 156 66 123
Oxford	497 156 170 63 113 74 82 109 260 173 115 102 135 361 152 470 354 49 60 80 212 188 523 217 171 126 164 96 153 230 253 41 144 104
Penzance	692 310 109 272 288 202 346 232 462 293 348 163 353 561 112 662 559 222 262 240 403 411 724 417 401 358 366 317 355 449 487 295 407 336 265
Perth	83 380 510 339 514 420 385 443 144 429 439 497 506 43 494 106 60 446 467 374 372 277 115 195 251 307 269 467 263 192 159 388 413 316 407 607
Plymouth	624 239 67 199 218 125 275 164 391 227 276 92 287 488 45 589 486 150 191 168 331 341 654 344 328 279 294 245 281 378 410 224 336 265 193 78 537
Preston	324 147 279 109 277 188 197 209 88 197 252 266 305 187 260 285 182 158 234 143 138 121 352 40 69 123 29 268 32 105 124 157 229 123 175 374 235 306
Salisbury	547 180 116 112 86 52 139 99 313 160 146 40 156 412 91 515 412 75 61 104 252 250 576 270 232 188 217 115 203 292 319 103 197 166 62 204 459 133 229
Sheffield	376 164 270 87 222 179 119 202 161 226 175 245 241 250 249 371 255 146 192 148 160 68 407 121 34 46 76 202 36 106 132 102 150 45 143 364 291 293 80 205
Shrewsbury	412 76 201 50 219 112 144 110 178 126 192 188 245 276 183 375 272 77 167 51 106 169 439 130 116 117 64 209 69 189 216 98 207 81 106 297 322 229 91 146 93
Southampton	569 202 139 128 63 75 132 123 339 181 140 54 145 437 114 547 436 99 50 130 280 253 599 235 193 241 104 224 297 108 191 171 67 228 483 157 251 23 205 174
Stoke	387 110 219 44 217 125 130 149 156 160 190 205 241 252 200 360 248 94 173 82 118 413 107 90 89 55 205 44 164 188 93 176 51 111 313 297 243 64 166 52 34 191
Stranraer	244 344 479 307 476 386 361 406 109 393 420 464 503 216 457 187 88 351 432 338 334 259 264 160 232 292 234 468 250 200 167 355 393 295 371 577 150 502 198 423 269 287 448 261
Taunton	554 166 126 152 50 203 87 318 149 228 41 281 415 32 515 412 124 95 258 269 581 270 256 209 221 179 208 306 335 152 264 193 121 147 462 78 231 66 221 152 89 170 429
York	325 200 313 108 268 221 153 241 117 249 210 287 281 191 291 325 208 193 162 94 83 144 185 86 185 406 238 340 83 247 60 144 252 119 228 263
LONDON	543 237 217 118 53 119 60 155 307 216 63 121 77 405 170 513 405 104 30 135 267 215 569 259 196 141 210 37 197 252 278 67 115 128 56 283 453 215 218 82 167 160 76 156 419 167 209

The following pages contain motorway maps and text as described on pages 6–8. In some cases more than one motorway will be found on a particular map and the page heading will only indicate the main one. To help avoid any confusion, we list below all the motorways with their relevant page numbers. The legend will help you to interpret the map sections and we also list a number of abbreviations which you might find in the text pages.

Legend

AA 15	AA road service centres. Breakdown and road service information. Normally 0900-1730 hrs
AA	AA service centres. Breakdown/information service normal hours
AA info	AA motorway information service centres. Normally 0900-1730 hrs. Callers only
AA 24 hour	AA service centres 24 hour breakdown/information service
3	Motorway with junction number (Letters may be used to aid identification in cases where junctions are not numbered)
	Motorway under construction
	Motorway projected
A3	Primary route
A35	Dual carriageway
A335	A road
B3036	B road
	Road under construction
6	Junction with restricted access
S	Service area
✈	Airport
6	Mileage between junctions and service areas.

Hotel Chains

Company	Abbreviations
Allied Hotels (Scotland)	AS
Anchor Hotels & Taverns Ltd	A
Ansells Brewery Co Ltd	An
Berni Inns Ltd	B
Best Western	BW
Charrington & Co	Ch
Comfort Hotels International	Co
Commonwealth Holiday Inns of Canada Ltd	Com
*Crest Hotels Ltd	Cr
De Vere Hotels Ltd	DeV
Embassy Hotels	E
Exec Hotels	Ex
Forest Dale Hotel Group	FD
Forum Hotels	F
Frederic Robinson Ltd	FR
Great British Hotels (Anchor, GW and Swallow Groups), See details under Anchor, Greenall Whitley and Swallow	
Greenall Whitley Hotels Ltd	GW
*Holiday Inn International	
Home Brewery Co Ltd	HB
Hotel Representative Incorporated for Berkeley, Claridges, Connaught, Savoy, Hyde Park, London; Royal Crescent, Bath, Avon; Lygon Arms, Broadway, Worcs; Grosvenor, Chester; Chewton Glen, New Milton, Hants	
Ind Coope (Alloa Brewery Co Ltd)	IC
Inter-Continental Hotels	ICon
Inter-Hotels	IH
Kingsmead Hotels Ltd	K
*Ladbroke Hotels Ltd	L
Minotels Ltd	M
*Mount Charlotte Hotels Ltd	MC
The North Hotels	N
Paten & Co Ltd	P
Percy R Brend (Hotels) Ltd	Br
Porter Group	Por
*Prestige Hotels	Pr
Prince of Wales Hotels Ltd	PW
Queens Moat House Ltd	Q
Rank Hotels Ltd	R
*Reo Stakis Organisation	St
Saxon Inn Motor Hotels Ltd	Sx
Scottish Highland Hotels	SH
Scottish & Newcastle Breweries Ltd	SN
*Swallow Hotels Ltd	Sw
*Thistle Hotels (Scottish & Newcastle Breweries)	T
Travco Hotels Ltd	Tr
Trusthouse Forte Hotels Ltd	THF
Wessex Taverns Ltd	WT
Whitbread Wessex (Retail)	W
Whitbread West Pennines Ltd	WWP
Wolverhampton & Dudley Breweries Ltd	W&D

Abbreviations

AM	Ancient Monument
Av	Avenue
FH	Farmhouse
GH	Guesthouse
La	Lane
NT	National Trust
NTS	National Trust for Scotland
PS	Picnic Site
R	Restricted junction – refer to direction panel opposite
Rd	Road
St	Street
Ter	Terrace

Garage franchises – see page 8

London - Milton Keynes [M1]

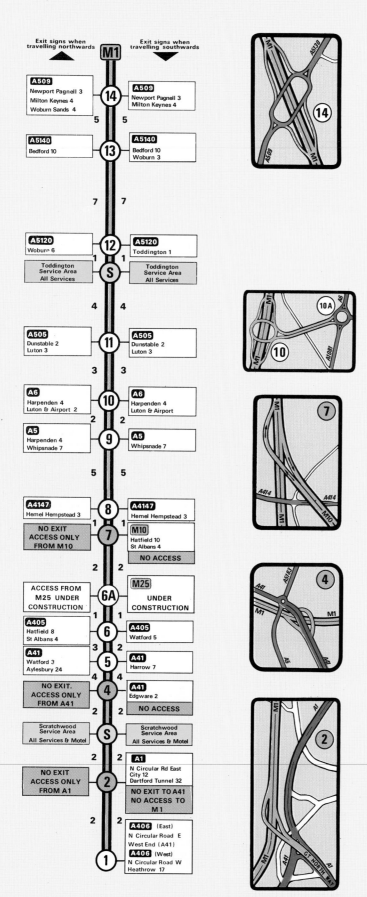

Exit signs when travelling northwards

Exit signs when travelling southwards

A509
Newport Pagnell 3
Milton Keynes 4
Woburn Sands 4
— (14) — **A509** Newport Pagnell 3 Milton Keynes 4

5 — 5

A5140
Bedford 10
— (13) — **A5140** Bedford 10 Woburn 3

7 — 7

A5120
Woburn 6
— (12) — **A5120** Toddington 1

Toddington Service Area All Services — (S) — **Toddington Service Area** All Services

4 — 4

A505
Dunstable 2
Luton 3
— (11) — **A505** Dunstable 2 Luton 3

3 — 3

A6
Harpenden 4
Luton & Airport 2
— (10) — **A6** Harpenden 4 Luton & Airport

2 — 2

A5
Harpenden 4
Whipsnade 7
— (9) — **A5** Whipsnade 7

5 — 5

A4147
Hemel Hempstead 3
— (8) — **A4147** Hemel Hempstead 3

NO EXIT ACCESS ONLY FROM M10 — (7) — **M10** Hatfield 10 St Albans 4 **NO ACCESS**

2 — 2

ACCESS FROM M25 UNDER CONSTRUCTION — (6A) — **M25 UNDER CONSTRUCTION**

1 — 1

A405
Hatfield 8
St Albans 4
— (6) — **A405** Watford 5

3 — 3

A41
Watford 3
Aylesbury 24
— (5) — **A41** Harrow 7

2 — 2

NO EXIT. ACCESS ONLY FROM A41 — (4) — **A41** Edgware 2 **NO ACCESS**

2 — 2

Scratchwood Service Area All Services & Motel — (S) — **Scratchwood Service Area** All Services & Motel

2 — 2

NO EXIT ACCESS ONLY FROM A1 — (2) — **A1** N Circular Rd East City 12 Dartford Tunnel 32 **NO EXIT TO A41 NO ACCESS TO M1**

2 — 2

— (1) — **A406** (East) N Circular Road E West End (A41) **A406** (West) N Circular Road W Heathrow 17

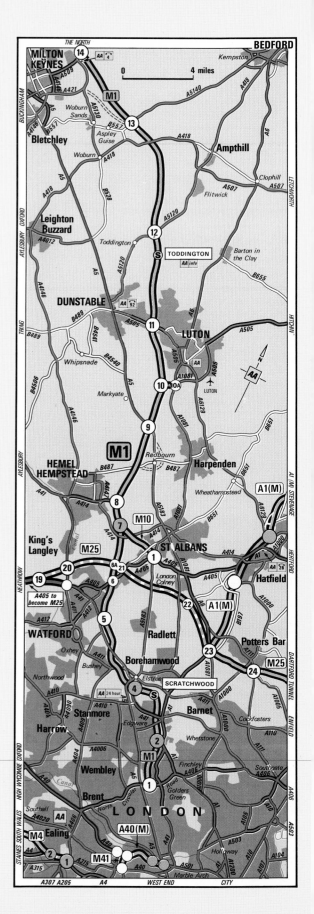

34

Junction 1

The London 'village' of Hampstead with its lovely Heath is nearby. To the west is Wembley Stadium.

WHERE TO STAY
★★★**Charles Bernard**
5 Frognal, Hampstead, NW3
☎01–794 0101
☆☆☆☆**Wembley International**
Empire Way, Wembley
☎01–902 8839
☆☆☆☆**BL Holiday Inn**
King Henry's Rd, Swiss Cottage NW3 (CH)
☎01–722 7711

WHERE TO EAT
××**Luigi's**
1–4 Belmont Pde, Finchley Rd, Golders Green NW11
☎01–455 0210
×**Hong**
30 Temple Fortune Pde, Finchley Rd, Golders Green NW11 ☎01–455 9444

GARAGES
Warwick Wright Motors
393 Edgware Rd, NW2 (Tal)
☎01–452 0041
Temple Fortune
1089 Finchley Rd, Golders Green, NW11 (Ren)
☎01–455 9935

Junction 2 R

There are a number of interesting places nearby including the R.A.F. Museum at Hendon.

WHERE TO STAY
☆☆☆☆**Hendon Hall**
Ashley Ln, Hendon NW4 (KH)
☎01–203 3341

WHERE TO EAT
Blue Angel
3 Long Ln, Finchley
☎01–349 4386

GARAGES
Aerodrome Autos
28 Aerodrome Rd, Hendon, NW4 (Cit)
☎01–203 0071
Northway Service Station
137 Great North Way, Hendon, NW4
☎01–203 3834

Scratchwood Services
(THF) ☎01–959 7395
Accommodation. Restaurant. 24hr cafeteria. Fast food. Shop. Petrol. Diesel. Breakdowns. Repairs. HGV Parking. Overnight parking for caravans £2.50. Baby-changing. For disabled: toilets, ramp. Credit cards – shop, garage, hotel, restaurants.

Junction 4 R

Just north of Elstree is a 185-acre country park beside the Aldenham reservoir.

WHERE TO STAY
★★★**Grim's Dyke Country Hotel**
Old Redding, Harrow Weald, Gt London ☎01–954 4227

WHERE TO EAT
Peking Duck
35 Broadway, Mill Hill, NW7
☎01–954 4050

All the fun of the fair comes to London's Hampstead Heath for the Late Summer Bank Holiday at the end of August.

GARAGES
Autoport
(Lap) Allum Ln, Elstree (Frd)
☎(0727) 50871
Stanmore Motor Co
65 Stanmore Hill, Stanmore (Tal Peu) ☎01–954 1699

Junction 5

The River Colne flows past the town of Watford which, although modernised, contains some interesting old buildings.

WHERE TO STAY
★★**Red Lion**
(Henekey's) Watling St, Radlett (THF) ☎(09276) 5341

WHERE TO EAT
×**Flower Drum**
16 Market St, Watford
☎(0923) 26711

GARAGES
Meriden Service Station
(Carport) York Way, Watford
☎(09273) 73607
Shaw & Kilburn
329 St Albans Rd, Watford
(Vau Opl Vlo) ☎(0923) 31716
Hadleigh Ross Autos
101–107 Sutton Rd, Watford (BL) ☎(09273) 79952

Junction 6

Abbots Langley was the birthplace of the only English pope, Nicholas Breakspear who became Hadrian IV some 800 years ago.

GARAGES
Chequers Service Station
Old Watford Rd, Bricket Wood ☎(09273) 72051
Sheepcot Service Station
(Blackaby & Pearce Motor Eng) North Orbital Rd, Watford ☎(09273) 74166

Junction 7 R

St Albans is a historic town with much to see including the cathedral and the

---

Roman remains of Verulamium.

Junction 8

Old and new mingle quite cheerfully in Hemel Hempstead where flower beds and canal-side walks border the town centre.

WHERE TO STAY
☆☆☆**Post House**
Breakspear Way, Hemel Hempstead (THF)
☎(0442)51121
☆☆☆**Hemel Hempstead Moat House**
London Rd, Bourne End, Hemel Hempstead (QM)
☎(04427) 71241
☆☆☆**Aubrey Park**
Hemel Hempstead Rd, Redbourne (BW)
☎(058285) 2105

WHERE TO EAT
Old Bell
High St, Hemel Hempstead
☎(0442) 52867

GARAGES
Adeyfield Recovery Service
1 St Albans Hill, Hemel Hempstead ☎(0442) 52515
Cupid Green Service Station
Redbourn Road, Hemel Hempstead ☎(0442) 52447
Marlowes
Hillfield Rd, Hemel Hempstead ☎(0442) 49494

Junction 9

To the west the Chiltern Hills become the Dunstable Downs, home of Whipsnade Zoo.

WHERE TO EAT
××**Willow Tree**
6 Church Green, Harpenden
☎(05827) 69358

GARAGES
F Ogglesby
17 Luton Rd, Harpenden (Vau) ☎(05827) 67776
Pinneys
Station Rd, Harpenden (Vlo)
☎(05827) 64311

M1 London–Milton Keynes   44 miles

From London's northern suburbs the M1 emerges to skirt the historic town of St Albans before continuing to the ultra-modern Milton Keynes.

Junction 10/10a

This is the junction for Luton Airport. Nearby is Luton Hoo, a country mansion by Robert Adam.

WHERE TO STAY
☆☆☆☆**Strathmore Thistle**
Arndale Centre, Luton (TS)
☎(0582) 34199
★★**Red Lion**
(Henekey) Castle St, Luton (THF) ☎(0582) 413881

GARAGES
Luton Airport Service Station
Eaton Green Rd, Luton
☎(0582) 30534
Motor Services
Unit 9C, Lye Trading Estate, Old Bedford Rd, Luton
☎(0582) 425104
Poynters Road
(Blackaby & Pearce) 182 Poynters Road, Luton (Fia)
☎(0582) 67742
Rix Manor
Leagrave Rd, Luton (BL)
☎(0582) 51221
Streatley
Barton Rd, Streatley, Luton
☎(0582) 591575
Sundon Park Service Station
(D Norris) Sundon Park Rd, Luton ☎(0582) 53008
Trimico
The Trading Estate, Chaul End, Luton (Frd)
☎(0582) 31133
Skyways Service Station
(CWA Motors) Luton
☎(0582) 419694

Junction 11

The motor trade towns of Luton and Dunstable are served by this junction.

WHERE TO STAY
☆☆☆☆**Chiltern**
Waller Av, Dunstable Rd, Luton (CRH) ☎(0582) 55911
☆☆☆**Crest-Luton**
Dunstable Rd, Luton (CRH)
☎(0582) 55955
★★★**Old Palace Lodge**
Church St, Dunstable
☎(0582) 62201
★★**Highwayman**
London Rd, Dunstable
☎(0582) 61999

Luton Hoo contains the Wernher Collection of art treasures.

GARAGES
B Evans Motors
202 High St South, Dunstable
☎(0582) 603473
Autocraft
Unit 9, Houghton Regis Ind. Estate, Cemetery Rd, Houghton Regis.
☎(0582) 602328

Toddington Services
(Granada) ☎(05255) 3881
Restaurant. Fast Food. HGV Cafeteria. Shop. Petrol, Diesel. Breakdowns. Repairs. HGV Parking. Long-term/overnight parking for caravans £3.50. Baby-changing. For disabled: toilets, ramp. Credit cards – shop, garage, restaurant. Footbridge. Road bridge.

Junction 12

Attractive countryside with pretty thatched cottages surrounds this junction.

WHERE TO EAT
×**French Horn**
Steppingley ☎(0525) 712051

GARAGES
One-O One
Ampthill Rd, Flitwick (BL)
☎(0525) 712244

Junction 13

The picturesque village of Woburn lies to the south. Woburn Abbey, a palatial 18th-century mansion has a 3000-acre park.

WHERE TO STAY
★★★**Bedford Arms**
George St, Woburn (KH)
☎(052525) 441
INN **Red Lion**
Wavendon Rd, Salford
☎(0908) 583117

WHERE TO EAT
Woburn Wine Lodge
13 Bedford St, Woburn
☎(052525) 439

GARAGES
Halt
(Guise Motor) Husborne Crawley ☎(0908) 583242
Gary Bolton
80–82 Station Rd, Woburn Sands ☎(0908) 582458
R G R
High St, Cranfield (Frd)
☎(0234) 750207
A White & Son
High St, Cranfield (BL Mgn)
☎(0234) 750205

Junction 14

The new town of Milton Keynes contains Europe's largest covered shopping centre and is the home of the Open University.

WHERE TO STAY
★★**Swan Revived**
High St, Newport Pagnell
☎(0908) 610565
★★**Woughton House**
Woughton on the Green, Milton Keynes
☎(0908) 661919

WHERE TO EAT
××**Glovers**
18–20 St John St, Newport Pagnell ☎(0908) 616398

GARAGES
7 to 12 Autos
7/8 Tower Cres, Neath Hill, Milton Keynes
☎(0908) 367293

LOCAL RADIO STATIONS

	Medium Wave		VHF/FM
	Metres	kHz	MHz
BBC Radio London	206	1458	94.9
IBA Capital Radio	194	1548	95.8
IBA LBC	261	1152	97.3
IBA Chiltern Radio			
Luton Area	362	828	97.6
Bedford Area	378	792	95.5

M1 Milton Keynes – Leicester   48 miles

Continuing northwards through the East Midlands, the M1 links with the M45 (for Coventry Airport), the M6 (for the North West) and the M69 (from Coventry). The historic battlefields of Naseby and Bosworth Field are passed.

At one time much of the country's heavy freight went by canal. Now the Grand Union Canal rests peacefully as the juggernauts thunder up and down the motorway.

Junction 14 – see page 35

Newport Pagnell Services
(THF) ☎(0908) 610142
24hr Cafeteria. Waitress-service 'Little Chef' (north side) 07.30–20.00.
Shop. Petrol. Diesel. Breakdowns. Repairs. Long-term/overnight parking for caravans £2.50. South side, reached via footbridge, has additional facilities: Accommodation, fast food, HGV parking, baby-changing, ramp and toilets for disabled. Credit cards – shop, garage, hotel, restaurants.

Junction 15

To the south, at Stoke Bruerne, is a fascinating Waterways Museum beside the Grand Union Canal. Stoke Park Pavilions, dating from the 17th century, can be seen nearby while a little farther afield are Towcester (horse-racing) and Silverstone (motor-racing).

WHERE TO STAY
★★★**Grand**
Gold St, Northampton
☎(0604) 34416
INN **Roadhouse**
16–18 High St, Roade
☎(0604) 863372

WHERE TO EAT
✕**Napoleon's Bistro**
9–11 Welford Road, Northampton
☎(0604) 713899
✕**Vineyard**
7 Derngate, Northampton
☎(0604) 33978
✕✕**French Partridge**
Horton ☎(0604) 870033
PS **Salcey Forest**
off B526 S of Northampton, 1m from Hartwell OS152 SP9798

GARAGES
Johns Motors
Letts Rd, Far Cotton, Northampton (Rel)
☎(0604) 64761
Mann Egerton & Co
Bedford Rd, Northampton (Ren RR) ☎(0604) 39645
W Grose
Queens Park Pde, Northampton (Vau Opl)
☎(0604) 712525
Henlys
Weedon Rd, Northampton (BL) ☎(0604) 54041
Kingsthorpe
42–50 Harborough Rd, Northampton (Aud MB VW)
☎(0604) 716716
Northampton Clutch Specialists
190 St Andrews Rd, Northampton ☎(0604) 32022
Wadham Stringer
Wellingborough Rd, Northampton (BL DJ LR)
☎(0604) 401141

Rothersthorpe Services
(Blue Boar) ☎(0604) 831885
Restaurant. Fast food. Shop. Vending machines. Petrol. Diesel. Breakdowns. Repairs. Long-term parking £5 (includes breakfast voucher). Baby-changing. For disabled: toilets, ramp **NB** steps in building. Credit cards accepted.

Junction 16

The 'boot and shoe' town of Northampton has much to interest visitors, including the 16th century Delapre Abbey, an interesting Industrial Museum and a milling museum housed in a former corn mill. Leisure facilities include Billing Aquadrome (a waterside park), Abington Park and Hunsbury Hill Country Park, centred on an Iron Age hill fort. To the north is Althorp, home of the Earl and Countess Spencer.

WHERE TO STAY
★★**Crossroads**
Weedon ☎(0327) 40354

WHERE TO EAT
PS **Harleston Heath**
1m S of Harleston on A428 Northampton – Rugby road OS152 SP7163

GARAGES
Total Service Centre
(Car & Truck Services)
Kislingbury ☎(0604) 830770
Bluecars Repairs
Litchborough Rd, Bugbrooke
☎(0604) 830010
Butts Hill
(J R Clayton Motors)
Bugbrooke ☎(0604) 830704
P J Green
81 High St, Flore (BL)
☎(0327) 40545
Freeways Motors
Watling St, Weedon
☎(0327) 40344
Stow Hill
(Clarke Bros) Weedon (W)
☎(0327) 40369
White Park Service Station
London Rd, Weedon
☎(0327) 830255

Airflow Streamlines
Hopping Hill, New Duston, Northampton (Frd)
☎(0604) 581121
Ash Tree Service Station
237–245 Main Rd, New Duston, Northampton
☎(0604) 830806

Watford Gap Services
(Blue Boar) ☎(03272) 5181
Restaurant. Picnic Area. Shop. Petrol. Diesel. Breakdowns. Repairs. Overnight parking for caravans £5 (includes breakfast voucher). Baby-changing. For disabled: toilets, ramp. Footbridge. Credit cards accepted.

Junction 17 R

The M45 Dunchurch Spur extends 8 miles west giving access to the A45 for Coventry Airport and southern Coventry. Dunchurch was the place where, on 5th November 1605, Guy Fawkes and his fellow conspirators awaited news from Westminster.

Junction 18

The town of Rugby with its famous school lies amidst delightful countryside. The Oxford canal, the River Avon and the River Swift are all nearby.

WHERE TO STAY
☆☆☆**Post House**
Crick (THF) ☎(0788) 822101
★★**Hillmorton Manor**
High Street, Hillmorton, Rugby ☎(0788) 65533
★★★**᠍Clifton Court**
Lilbourne Rd, Clifton upon Dunsmore ☎(0788) 65033
GH **Grosvenor House**
81 Clifton Rd, Rugby
☎(0788) 3437

GARAGES
Halfway
Crick Cross Rds, Kilsby (Fia)
☎(0788) 822226

Leicester's Gothic Clock Tower commemorates four benefactors of the city.

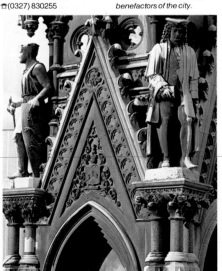

D Shorten Motors
A5/M1 Link Rd, Kilsby
☎(0788) 822988
Hillmorton
102 Hillmorton Rd, The Parade, Hillmorton
☎(0788) 2515

Junction 19 R

The M6 branches off here heading north west for Coventry, Birmingham and eventually Carlisle. See pages 64–73 for details.

Junction 20

The small town of Lutterworth lies alongside the motorway here – the place where John Wycliffe translated the Bible into English in the 14th century and where the jet engine was developed in 1941. South-east at Swinford is Stanford Hall and at Naseby is a museum relating the battle of 1645.

WHERE TO STAY
★★**Moorbarns**
Watling St, Lutterworth
☎(04555) 2237

GARAGES
Broughton Astley Motors
Unit 6, Oaks End Estate, Gilmorton Rd, Lutterworth
☎(04555) 56453
Burtons
Church St, Lutterworth (Frd)
☎(04555) 2363
Wycliffe
(J Richardson & Sons)
Bitteswell Rd, Lutterworth (Tal Peu) ☎(04555) 2177

Junction 21

The M69 links Leicester to Coventry – see pages 84–85 for details. Leicester is a prosperous and interesting city on the River Soar. Shopping facilities are particularly good and of the many museums the Museum of Technology is well worth a visit.

WHERE TO STAY
☆☆☆**Leicester Forest Moat House**
Hinkley Rd, Leicester Forest East (QM) ☎(0533) 394661
☆☆☆**Post House**
Braunstone Lane East, Leicester (THF)
☎(0533) 896688
★★★★**Grand**
Granby St, Leicester (EH)
☎(0533) 555599
☆☆☆**Holiday Inn**
St Nicholas Circle, Leicester
☎(0533) 531161
★★★**Belmont**
De Montfort St, Leicester (BW) ☎(0533) 544773
★★**Charnwood**
48 Leicester Rd, Narborough
☎(0533) 862218

GH **Forest Lodge**
Desford Rd, Kirby Muxloe
☎(0533) 393125
GH **Old Tudor Rectory**
Main St, Glenfield
☎(0533) 312214
GH **Alexandra**
342 London Rd, Stoneygate, Leicester ☎(0533) 703056
GH **Daval**
292 London Rd, Leicester
☎(0533) 708234
GH **Stanfre House**
265 London Rd, Leicester
☎(0533) 704294
★**Rowans**
290 London Rd, Leicester
☎(0533) 705364
GH **Burlington**
Elmfield Av, Leicester
☎(0533) 705112
GH **Scotia**
10 Westcotes Dr, Leicester
☎(0533) 549200

WHERE TO EAT
✕✕✕**Manor**
Glen Parva Manor, The Ford, Little Glen Rd, Glen Parva
☎(0533) 774604
Du Cann's Wine Bar
29 Market St, Leicester
☎(0533) 556877
Good Earth
19 Free Ln, Leicester
☎(0533) 26260
Spanish Place
38A Belvoir St, Leicester
☎(0533) 542830

GARAGES
Continental Motors
34–36 Western Rd, Leicester
☎(0533) 548249
Forum Service Centre
197 Narborough Rd, Leicester ☎(0533) 545135
Jordon Motors
30–34 Narborough Rd South, Leicester (Vlo)
☎(0533) 895952
Mann Egerton
Welford Rd, Leicester (RR Ren) ☎(0533) 548757
Hanger Motors
Welford Rd, Leicester (Frd)
☎(0533) 706215
A1 Autos
77–79 Church Gate, Leicester ☎(0533) 27459
Bishop & Bishop
Leicester, Wigston (BL Tal Peu) ☎(0533) 881601
Moat Service Station
Bull Head St, Wigston
☎(0533) 883796

Leicester Forest East Services
(Welcome Break)
☎(0533) 386801
24hr Cafeteria. Waitress-Service restaurant. Hostess-service Butteries. Fast food. HGV cafe. Shop. Vending machines. Playground. Petrol. Diesel. Liquid Petroleum Gas. Breakdowns. HGV parking. Long-term/overnight parking for caravans £3.00. Baby-changing. For disabled: toilets, lift. Footbridge. Road bridge.

LOCAL RADIO STATIONS

	Medium Wave		VHF/FM
	Metres	*kHz*	*MHz*
BBC Radio Northampton	271	1107	96.6
BBC Radio Leicester	358	837	95.1
IBA Leicester	238	1260	97.1
IBA Northampton station due to open late '84 or early '85			

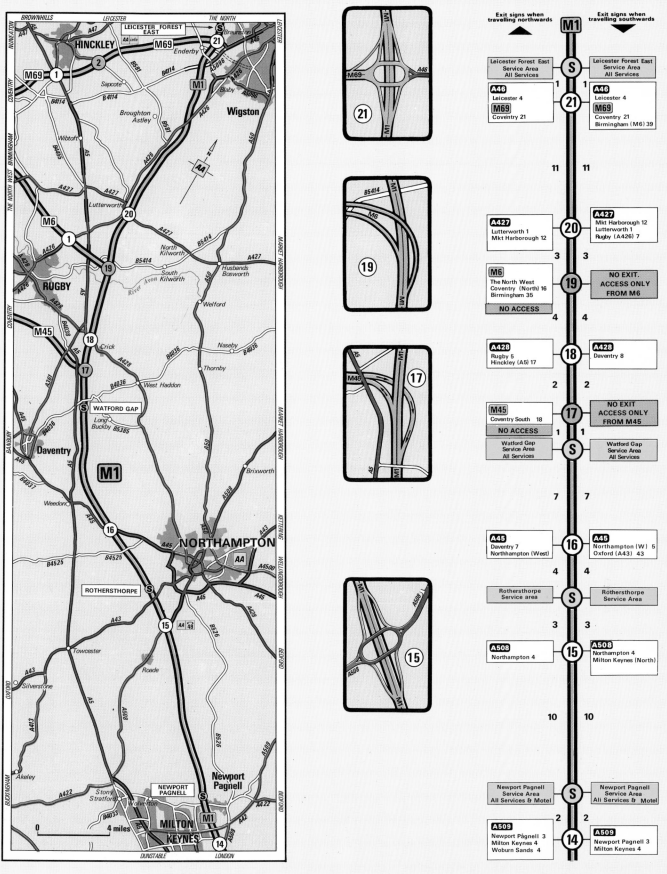

Leicester - Bolsover M1

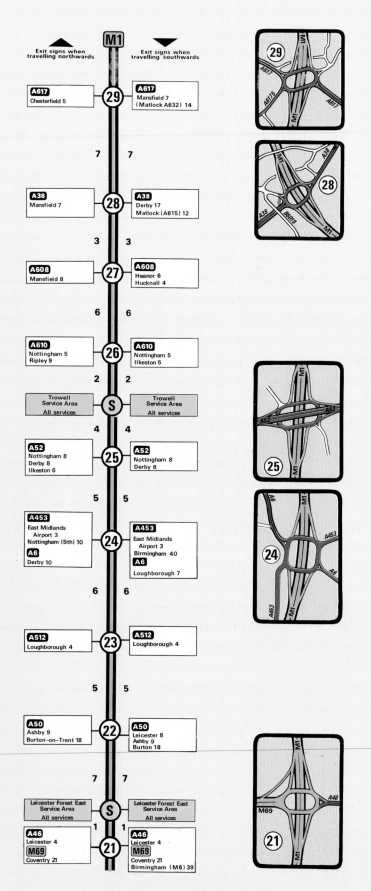

Exit signs when travelling northwards

Exit signs when travelling southwards

M1

Northbound	Jct	Southbound
A617 Chesterfield 5	29	**A617** Mansfield 7 (Matlock A632) 14
	7 / 7	
A38 Mansfield 7	28	**A38** Derby 17 Matlock (A615) 12
	3 / 3	
A608 Mansfield 8	27	**A608** Heanor 6 Hucknall 4
	6 / 6	
A610 Nottingham 5 Ripley 9	26	**A610** Nottingham 5 Ilkeston 5
	2 / 2	
Trowell Service Area All services	S	Trowell Service Area All services
	4 / 4	
A52 Nottingham 8 Derby 8 Ilkeston 6	25	**A52** Nottingham 8 Derby 8
	5 / 5	
A453 East Midlands Airport 3 Nottingham (Sth) 10 **A6** Derby 10	24	**A453** East Midlands Airport 3 Birmingham 40 **A6** Loughborough 7
	6 / 6	
A512 Loughborough 4	23	**A512** Loughborough 4
	5 / 5	
A50 Ashby 9 Burton-on-Trent 18	22	**A50** Leicester 8 Ashby 9 Burton 18
	7 / 7	
Leicester Forest East Service Area All services	S	Leicester Forest East Service Area All services
	1 / 1	
A46 Leicester 4 **M69** Coventry 21	21	**A46** Leicester 4 **M69** Coventry 21 Birmingham (M6) 39

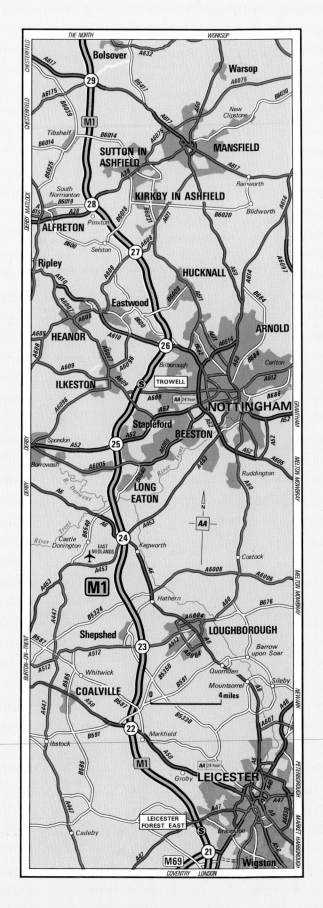

Junction 21 and Leicester Forest East Services – see page 36

The River Derwent flows through some magnificent scenery from the Peak District to the Derbyshire Dales. Here it is shown near Cromford, west of junction 28.

Junction 22

To the east 1,223 acres of the Charnwood Forest have been turned into the Bradgate Park and Swithland Woods Country Park. On the edge of Coalville, Donington-le-Heath Manor is a well preserved medieval house.

WHERE TO STAY
★★**Johnscliffe**
73 Main St, Newtown Linford (EXEC) ☎(0530) 242228

GARAGES
Flying Horse
5–7 Shaw Ln, Markfield ☎(0530) 244361

Junction 23

Loughborough is a manufacturing town with a modern University of Technology. The Great Central Railway is a private steam railway running 5 miles to Rothley.

WHERE TO STAY
★★★**King's Head**
High St, Loughborough (EH) ☎(0509) 214893
★★**Cedars**
Cedar Rd, Loughborough ☎(0509) 214459
★★**Great Central**
Great Central Rd, Loughborough (MINO) ☎(0509) 263405
GH **De Montfort**
88 Leicester Rd, Loughborough ☎(0509) 216061
GH **Sunnyside**
The Coneries, Loughborough ☎(0509) 216217
FH Miss F White **Talbot House** Thringstone ☎(0530) 222233

WHERE TO EAT
××**Cottage in the Wood**
Maplewell Rd, Woodhouse Eaves ☎(0509) 890318
××**Harlequin**
11 Swan St, Loughborough ☎(0509) 215235

PS **Nanpantan Outwoods**
1½m W of Woodhouse Eaves off B591 OS129 SK5114
PS **Beacon Hill**
1½m W of Woodhouse Eaves off B591 OS129 SK5114

GARAGES
Poyser & Rutherford
67 Brook St, Shepshed (Frd) ☎(0509) 503207
L E Jackson
(Coachworks) Queens Rd, Loughborough ☎(0509) 30811
Mann Egerton
Wood Gate, Loughborough (BL DJ) ☎(0509) 262710
Southfield Service Station
(MCG Transport) Southfield Rd, Loughborough ☎(0509) 212330
W S Yeates
Derby Rd, Loughborough (Vlo) ☎(0509) 217777

Junction 24

Leave the motorway here for East Midlands airport, or if you are heading for the motor racing at Donington Park. Melbourne Hall, the stately home of the Marquess of Lothian, is nearby.

WHERE TO STAY
★★★**Yew Lodge**
33 Packington Hill, Kegworth ☎(0509) 2518
★★**Donington Manor**
High St, Castle Donington ☎(0332) 810253
★★★**Priest House Inn**
Kings Mills, Castle Donington ☎(0332) 810649
GH **Delven**
12 Delven Ln, Castle Donington ☎(0332) 810153

WHERE TO EAT
××**Lady in Grey**
Wilne Ln, Shardlow ☎(0332) 792331

GARAGES
Airport Service Station
(Western Car Co) Castle Donington ☎(0332) 810710

Smiths
The Green, Diseworth ☎(0332) 810467

Junction 25

Nottingham, famous for its fine lace and Robin Hood, also attracts cricket fans to Trent Bridge and water sports enthusiasts to Holme Pierrepont

WHERE TO STAY
☆☆☆**Post House**
Bostocks Ln, Sandiacre (THF) ☎(0602) 397800
☆☆☆**Novotel Nottingham/Derby**
Bostock Ln, Long Eaton ☎(06076) 60106
★★**Europa**
Derby Rd, Long Eaton ☎(0602) 728481
GH **Camden**
85 Nottingham Rd, Long Eaton ☎(06076) 62901

An industrial steam locomotive, now on show at Loughborough's Great Central Railway.

M1 Leicester–Bolsover   46 miles

Through the Charnwood Forest and on towards Derbyshire, the motorway passes close by the East Midlands airport on the way.

GH **Brackley House**
31 Elm Av, Beeston ☎(0602) 251787

WHERE TO EAT
××**Wilmott Arms**
Borrowash ☎(0332) 672222
Grange Farm Restaurant
Toton ☎(06076) 69426

GARAGES
Risley
81 Derby Rd, Risley ☎(0602) 397280
Hemlockstone
Hickings Ln, Stapleford (Rel) ☎(0602) 252525
Sandicliffe
Nottingham Rd, Stapleford (Frd) ☎(0602) 395000
Toton Lane
Toton Ln, Stapleford ☎(0602) 392452
V Albion
21 Pasture Rd, Stapleford (Peu Tal) ☎(0602) 394444
Draycott
Victoria Rd, Draycott ☎(03317) 2359
Kennings
42 Nottingham Rd, Long Eaton (BL MG) ☎(06076) 2143
Bramcote Hills
57 Derby Rd, Beeston ☎(0602) 254206
Ambercote Service Station
(XL Motors)
Bramcote Ln, Nottingham ☎(0602) 256422

Trowell Services
(Granada) ☎(0602) 320291
Restaurant. Fast food. Shop. Playground. Petrol. Diesel. Breakdowns. Repairs. HGV parking. Long-term/overnight parking for caravans £3.50. Baby-changing. For disabled: toilets, lift, ramp. Footbridge. Credit cards – shop, garage, restaurant.

This former Glasgow tram is now among the many exhibits at the Crich Tramway Museum.

Junction 26

Sherwood Forest stretches away to the north of Nottingham, although it is no longer the dense oak woodland we all imagine from the Robin Hood stories.

GARAGES
NCV
Hucknall Rd, Nottingham (BL) ☎(0602) 272915
Sheppard Wood Service Station
Bracebridge Dr, Nottingham ☎(0602) 293454
Main Road
(Rod Blanchard Motors) Watnall (Sub Jen) ☎(0602) 382781

Junction 27

Newstead Abbey to the east, rebuilt as a house in the 16th and 17th centuries, was the home of Lord Byron. It lies amidst attractive wooded countryside.

Junction 28

Mansfield is an industrial town in a coal-mining area. To the west at Crich, is a unique Tramway museum on the lovely Derwent Valley.

Junction 29

At the heart of the Dukeries, this junction is handy for visiting Hardwick Hall just a short distance to the south. To the north-west is Chesterfield with its famous twisted spire.

WHERE TO STAY
☆☆☆☆**Swallow**
Carter Ln East, South Normanton (SW) ☎(0773) 812000

GARAGES
Carnfield
(A Kettle & Son) Alfreton Rd, South Normanton (Frd Sko Yam Suz Ves) ☎(0773) 811251
Fourways
(D & M A Bacon) Kirkby Ln, Pinxton ☎(0773) 810694
Kirkby Folly Service Station
Kirkby Folly Rd, Sutton-in-Ashfield ☎(0623) 555264

GARAGES
R Staley & Son
Mansfield Rd, Glapwell (Tal) ☎(0623) 810634
Woodleigh Motor Sales
North Wingfield Rd, Grassmoor (AR Fia) ☎(0246) 850686

LOCAL RADIO STATIONS

	Medium Wave		VHF/FM
	Metres	kHz	MHz
BBC Radio Leicester	358	837	95.1
BBC Radio Nottingham	197	1521	95.4
Mansfield area	189	1584	
IBA Leicester	238	1260	97.1
IBA Radio Trent	301	999	96.2

M1 Bolsover–Leeds   51½ miles

Into industrial South Yorkshire, the motorway cuts through the outskirts of Sheffield before crossing the coalfields and coming to an end in Leeds.

Junction 29 – see page 39

Junction 30

At Whittington, north of Chesterfield is Revolution House where the plan was hatched to invite William of Orange to become King of England.

GARAGES
Bridgehouse
(Penmole), Renishaw Hill, Renishaw ☎(0246) 810600
Hargreaves Industrial Services
Station Rd, Clowne
☎(0246) 810215
P Blake
Chesterfield Rd, Staveley
(Opl Vau Ren)
☎(0246) 863008

Woodall Services
(THF) ☎(0742) 486434
24hr cafeteria (south side). Waitress service 'Little Chef'. Shop. Petrol. Diesel. Breakdowns. Repairs. HGV parking. Long-term/overnight parking for caravans £2. Baby-changing. For disabled: toilets, ramp. Footbridge. Credit cards – shop, garage, restaurants.

Junction 31

Amidst overspill development on Sheffield's south-eastern side is the new Rother Valley Country Park, still very much in the process of becoming a leisure area. It will eventually provide all kinds of sports facilities.

WHERE TO EAT
××**Red Lion**
Todwick ☎(0909) 771654
××**Deans**
Falcon Square, Dinnington
☎(0909) 562455

GARAGES
Auto Recovery and Repair Service
28/34 Main St, Aughton
☎(0742) 873146

Junction 32

The M18 branches off here towards Doncaster and The Humber Estuary – see page 92 for details. To the East, Roche Abbey, which dates from 1147, has some walls still standing to their full height.

Junction 33/34

Here is an industrial part of Sheffield where steelworks and a maze of railway sidings lie between the rivers Rother and Don. Wincobank Hill, west of junction 4 is topped with an Iron-age hill fort. Rotherham lies to the east.

A tranquil corner of the Cannon Hall Country Park at Cawthorne.

WHERE TO STAY
☆☆☆☆**Carlton Park**
Moorgate Rd, Rotherham
☎(0709) 64902
★★**Brentwood**
Moorgate Rd, Rotherham
☎(0709) 2772
★**Elton**
Main St, Bramley, Rotherham
☎(0709) 545681
★★★**Royal Victoria**
Victoria Station Rd, Sheffield
☎(0742) 78822
★★**St Andrew's**
46–48 Kenwood Rd, Sheffield (SW)
☎(0742) 550309
★★★★**Grosvenor House**
Charter Square, Sheffield (THF) ☎(0742) 20041
★★★**St George**
Kenwood Rd, Sheffield (SW)
☎(0742) 583811
★★**Rutland**
452 Glossop Rd, Broomhill, Sheffield (Inter-Hotels)
☎(0742) 665215
GH **Sharrow View**
13 Sharrow View, Nether Edge, Sheffield
☎(0742) 51542

WHERE TO EAT
××**Dore**
Church Ln, Sheffield
☎(0742) 365948
Gangsters
11–17 Division St, Sheffield
☎(0742) 22861
Nameless
16–18 Cambridge St, Sheffield ☎(0742) 29751
Raffles
Charles St, Sheffield
☎(0742) 24921

GARAGES
Kirkby Central
128 Wellgate, Rotherham (Vau Opl) ☎(0709) 75571
Moorhouse & Alstead
132/8 Fitzwilliam Rd, Rotherham (Peu)
☎(0709) 2213

Rotherham Toyota Centre
7–9 Canklow Rd, Rotherham
☎(0709) 60681
Abbey Motors
175 Rutland Rd, Pitsmoor, Sheffield ☎(0742) 369041
Autoways
Brown St, Sheffield (Frd)
☎(0742) 78271
Bramhall
1 Saville St, Sheffield (Vau)
☎(0742) 751565
E W Hatfield Ltd
100 Corporation St, Sheffield (DJ LR BL Rar)
☎(0742) 730291
Kennings
Tenter St, Sheffield (BL RR)
☎(0742) 71141

Junction 35

Sheffield's northern suburbs give way to wood and moorland to the west of this junction.

GARAGES
Woolley Wood Service Station
251 Ecclesfield Rd, Ecclesfield, Sheffield
☎(0742) 467919
Fitzwilliam Service Station
The Common, Ecclesfield
☎(0742) 468706

Junction 36

In the midst of the South Yorkshire coalfields are two country parks – Westwood, to the south-west and Worsborough Mill to the north. Within the latter is an interesting museum centred on a working watermill.

GARAGES
Diggles
Barnsley Rd, Wombwell (Frd)
☎(0226) 752374

Junction 37

The coal capital of Barnsley is to the east. At Cawthorne, Cannon Hall is an 18th-century house containing a museum and collection of paintings.

WHERE TO STAY
★★**Queens**
Regent St, Barnsley (AHT)
☎(0226) 84192
★**Royal**
Church St, Barnsley (AHT)
☎(0226) 203658
★★★**Ardsley House**
Doncaster Rd, Ardsley (BW)
☎(0226) 89401

WHERE TO EAT
××**Brooklands**
Barnsley Rd, Dodworth
☎(0226) 6364
PS **Cannon Hall Park**
½m N of Cawthorne off A635
OS110 SE2708

GARAGES
Highway Recovery
Recovery House, Sheffield Rd, Barnsley ☎(0226) 88288
Lookers
New St, Barnsley (Vau Opl)
☎(0226) 89181
Manor Flint Service Station
Wombwell Ln, Stairfoot, Barnsley ☎(0226) 5024
Worsborough Bridge
(L V Grimes & Sons) Barnsley (Opl Vau) ☎(0226) 203552

Junction 38

Almost adjacent to this junction is the Bretton Country Park, its large lakes watered by the River Dearne.

Woolley Edge Services
(Granada) ☎(0924) 85371
Restaurant 24hrs summer & weekends, 07.00–19.00 or 21.00 other times. Shop. Petrol. Diesel. Breakdowns. Repairs. Long-term/overnight parking for caravans £3.50. Baby-changing. For disabled: toilets. Credit Cards shop, garage, restaurant.

Junction 39

The motorway crosses the River Calder here and just to the east is the site of the Battle of Wakefield in 1460 where Richard,

Duke of York fell to the Lancastrians.

WHERE TO STAY
★**Sandel Court**
108 Barnsley Rd, Wakefield
☎(0924) 258725
★★★**Walton Hall**
Walton, Wakefield
☎(0924) 257911

GARAGES
Arnold G Wilson
Doncaster Rd, Wakefield (LR BL DJ Rar) ☎(0924) 377261

Junction 40

The motorway passes between Ossett and Wakefield here – towns which prospered on the cloth trade.

WHERE TO STAY
☆☆☆☆**Post House**
Queen's Dr, Ossett, Wakefield (THF)
☎(0924) 276388
★★★**Stoneleigh**
Doncaster Rd, Wakefield
☎(0924) 369461
★★★**Swallow**
Queens St, Wakefield (SW)
☎(0924) 372111

WHERE TO EAT
Venus Restaurant
51 Westgate, Wakefield
☎(0924) 75378

GARAGES
Gledhill
Kingsway, Ossett (BL)
☎(0924) 274253
Appleyard
Ings Rd, Wakefield (BL)
☎(0924) 70100
Glanfield Lawrence
68 Ings Rd, Wakefield (Vau Opl) ☎(0924) 372812
Wensleys
68 Ings Rd, Wakefield (VW Aud) ☎(0924) 375588

Junction 41

A number of small communities lie between Wakefield and Leeds – Ardsley, Outwood and Stanley can be reached from this junction.

GARAGES
Heybeck
Leeds Rd, Woodkirk
☎(0924) 472660

Junction 42 (M62)

By joining the M62 here it is possible to reach Liverpool in the west or Hull to the east – see pages 86–93 for details.

Junction 43

Beyond the River Aire to the north east is Temple Newsam House and Park, a splendid Tudor and Jacobean house standing

in 900 acres on the outskirts of Leeds.

WHERE TO STAY
☆☆☆**Stakis Windmill**
Mill Green View, Seacroft, Leeds (SO) ☎(0532) 732323

Junctions 44–47 R

Any of these junctions, some of which are restricted, are within reach of the places listed below. They approach the centre of Leeds, not only a major commercial city, but also a centre for theatrical, musical and sporting events.

WHERE TO STAY
☆☆☆☆**Ladbroke Dragonara**
Neville St, Leeds (LB)
☎(0532) 442000
★★★★**Queen's**
City Square, Leeds
☎(0532) 431323
★★★**Metropole**
King St, Leeds (THF)
☎(0532) 450841
★★★**Merrion**
Merrion Centre, Leeds (KH)
☎(0532) 439191
★**Hartrigg**
Shire Oak Rd, Headingley
☎(0532) 751568
GH **Budapest**
14 Cardigan Rd, Headingley
☎(0532) 756637
GH **Clock**
317 Roundhay Rd, Gipton Wood ☎(0532) 490304
GH **Highfield**
79 Cardigan Rd, Headingley
☎(0532) 752193
GH **Oak Villa**
57 Cardigan Rd, Headingley
☎(0532) 758439
GH **Trafford House**
18 Cardigan Rd, Headingley
☎(0532) 752034
GH **Aragon**
250 Stainbeck Ln, Meanwood (Minotel) ☎(0532) 759306

WHERE TO EAT
×××**Gardini's Terrazza**
Minerva House, 16 Greek St, Leeds ☎(0532) 432880
××**New Milano**
621 Roundhay Rd, Leeds
☎(0532) 659752

GARAGES
Wallace Arnold Sales & Service
123 Hunslet Rd, Leeds (Vau Opl) ☎(0532) 439911
KP Motor Cycles
No 1 The Calls, Leeds (Hon Yam Suz) ☎(0532) 460705
Ringways
Whitehall Rd, Leeds (Frd)
☎(0532) 634222
International Auto Safety Centre
Armley Rd, Leeds
☎(0532) 39776
Churwell Service Station
Elland Rd, Leeds
☎(0532) 716706
Grandstand Service Station
247 Elland Rd, Leeds
☎(0532) 712023
Inglewood Service Station
York Rd, Leeds
☎(0532) 648010
A G Wilson
495 Harrogate Rd, Alwoodley, Leeds (BL)
☎(0532) 684391
Appleyard
Roseville Rd, Leeds (BL DJ RR) ☎(0532) 432731
Tricentrol Cars
Roundhay Rd, Leeds (Frd)
☎(0532) 455955
Kirkstall Service Station
Kirkstall Ln, Leeds
☎(0532) 784178

LOCAL RADIO STATIONS			
	Medium Wave		**VHF/FM**
	Metres	*kHz*	*MHz*
BBC Radio Sheffield	290	1035	97.4
Sheffield City area	290	1035	88.6
BBC Radio Leeds	388	774	92.4
IBA Radio Hallam			
Sheffield area	194	1548	95.2
Rotherham area	194	1548	95.9
IBA Radio Aire	362	828	94.6

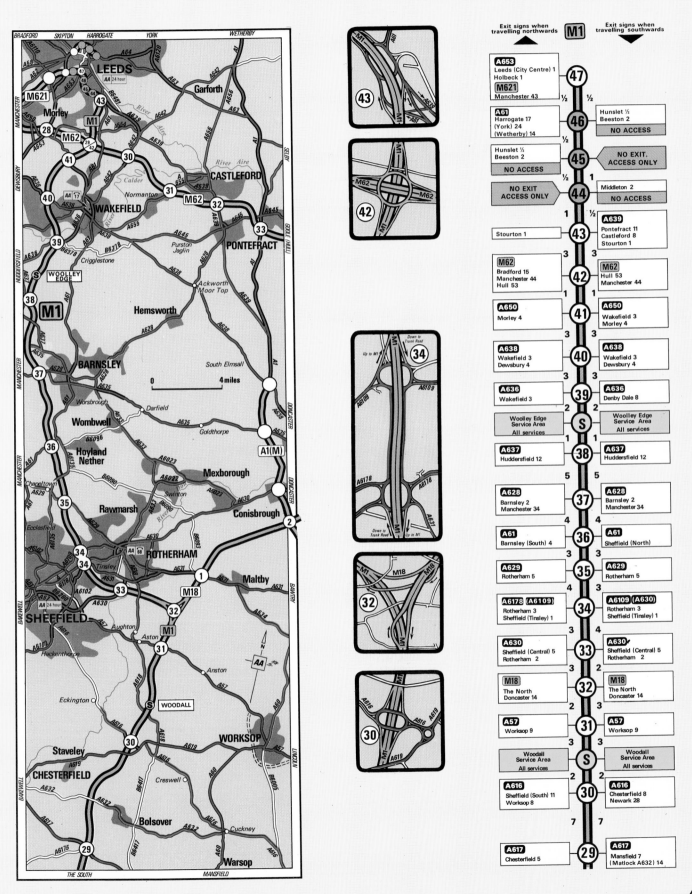

Rochester - Gillingham M2

Swanley - Hollingbourne M20

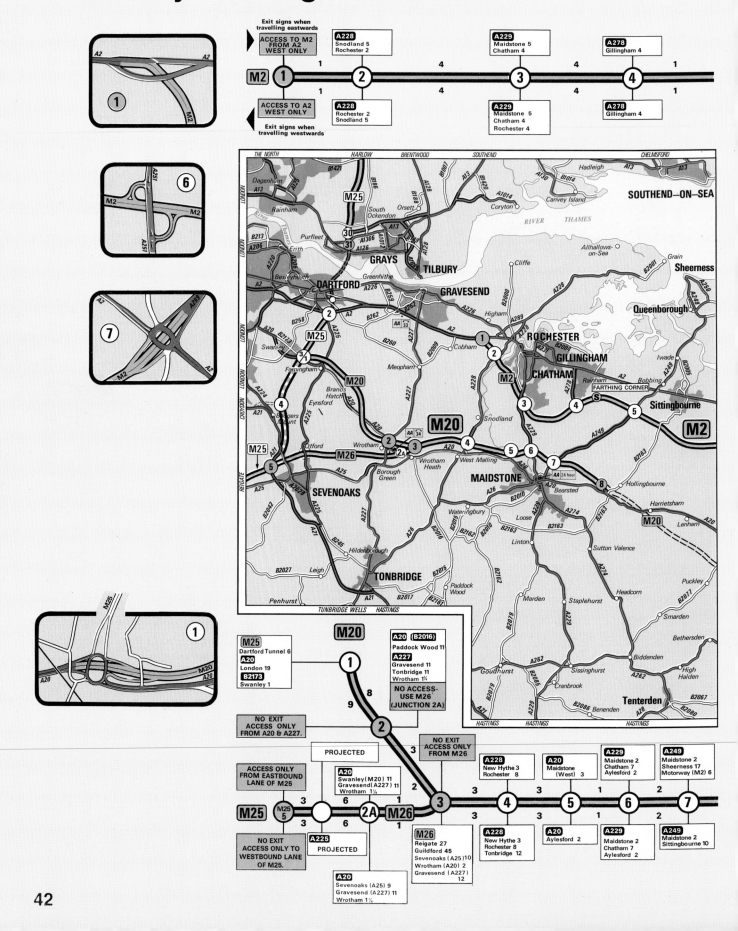

Rochester's cathedral, dating mainly from the 12th to 14th centuries, stands in a pleasant Close, once the site of a Benedictine monastery. The religious history of this Medway town in fact stretches back as far as the early part of the 7th century when a Saxon church stood on the site.

M2 Rochester – Gillingham 9 miles

Bypassing the Medway towns of Rochester, Chatham and Gillingham, the motorway follows the line of the North Downs.

houses are open to the public – Owletts, dating from the 17th century, and 16th century Cobham Hall, now a girls public school.

WHERE TO STAY
☆☆☆ **Inn on the Lake**
Watling St (A2), Shorne
☎(047482) 3333
GH **Greystones**
25 Watts Way, Rochester
☎(0634) 47545

WHERE TO EAT
PS **Camer Park**
½m N of Meopham off A227
OS177 TQ6566
PS **Shepherds Gate Transit Picnic Site**
3½m N of Rochester At B2009 interchange, OS177 TQ6869

Junction 1 R

The motorway links with the A2 from Dartford here and at nearby Cobham two

GARAGES
Rochester Motor Co
High St, Rochester (Tal Peu)
☎(0634) 42231
Vines
18 Maidstone Rd, Rochester
☎(0634) 41344
WJF Motors
2 Dunnings Ln, Rochester
☎(0634) 409288

Junction 3

Crossing the Medway, the motorway runs to the south of Chatham, a major naval town from the time of Henry VIII until the recent closure of the dockyard.

WHERE TO STAY
☆☆☆ **Crest**
Maidstone Rd, Rochester
(CRH) ☎(0634) 687111

WHERE TO EAT
✕✕ **Hengist**
7–9 High St, Aylesford
☎(0622) 79273

GARAGES
Medway
(Toyota) Airport Ind Est, Maidstone Rd, Rochester
(Toy) ☎(0634) 49247

Junction 2

The ancient city of Rochester on the west bank of the Medway has two castles – Rochester and Upnor – and a 12th- to 14th-century cathedral. The fine old buildings of the city have many Dickensian associations.

Robin Hood Service Station
Robin Hood Ln, Walderslade
☎(0634) 61309
Dutton Forshaw
20 Meadway St, Chatham (BL DJ) ☎(0634) 41122
Medspray
6 Second Av, Chatham
☎(0634) 47670

Junction 4

Adjacent to Chatham is Gillingham, the largest of this trio of towns and home of the Royal Engineers. An interesting museum tracing the history of the Regiment can be visited at Brompton Barracks.

WHERE TO STAY
★ **Park**
Nelson Rd, Gillingham
☎(0634) 51546

GARAGES
Greens
London Rd, Rainham (Vau Opl) ☎(0634) 31242
Autoyachts
171 Pier Rd, Gillingham (Fia Lnc) ☎(0634) 52333
Highland
(P Heard Services) Detling
☎(0622) 39864

Junction 1

The M20 links with the M25 London Orbital Motorway here – see pages 74–77 for details. To the south at Eynsford is the ruined Eynsford Castle, the mainly 18th-century Lullingstone Castle and the excavated Roman Villa. Brands Hatch motor racing circuit is a little way to the east.

WHERE TO EAT
PS **North Cray Transit Picnic Site**
1m N of Ruxley Corner (A20) OS177 TQ4871

GARAGES
Birchwood Motor Works
Birchwood Corner, Swanley
☎(0322) 63448
Dawes
Station Rd, Swanley
☎(0322) 62211

Junctions 2/3 and Junction 2A of M26 R

These junctions are all close together around Wrotham. Within easy reach of all three are the lovely Great Comp Gardens at Borough Green and Old Soar Manor at Plaxtol.

WHERE TO STAY
Post House
(at junction M26/M20/A20) Wrotham Heath (THF)
☎01–567 3444

GARAGES
Tower
(Valley Autos) Wrotham Hill, Wrotham ☎(0732) 822357

Borough Green
Maidstone Rd, Borough Green ☎(0732) 882017
Addington Motors
London Rd, Addington
☎(0732) 841839

Junction 4

Southwards at West Malling is St Leonards Tower, the surviving part of a former castle.

WHERE TO EAT
☆☆☆ **Larkfield**
London Rd, Larkfield (AHT)
☎(0732) 846858

WHERE TO EAT
✕✕✕ **Wealden Hall**
773 London Rd, Larkfield
☎(0732) 840259

GARAGES
Ham Hill
(Sparkdrill) Malling Rd, Snodland ☎(0634) 24204

Junction 5

Following the River Medway towards Maidstone, the motorway passes south of Aylesford Priory, a restored 13th- to 14th-century Carmelite House. Allington Castle is to the south on the bank of the river.

WHERE TO EAT
✕✕ **Hengist**
7–9 High Street, Aylesford
☎(0622) 79273

GARAGES
Tonbridge Road Garage
Tonbridge Rd, Barming
☎(0622) 29638

Junction 6

Maidstone is the bustling county town of Kent, a historic town as well as a modern centre for the surrounding agricultural area.

WHERE TO STAY
★★ **Royal Star**
High St, Maidstone (EH)
☎(0622) 55721

WHERE TO EAT
✕✕ **Running Horse**
Sandling, Maidstone
☎(0622) 52975
Pilgrim's Halt
98 High St, Maidstone
☎(0622) 57281

GARAGES
Boxley Rd Service Station
Boxley Rd, Maidstone
☎(0622) 52724
Chatham Rd Service Station
(Gatward) Chatham Rd, Maidstone ☎(0622) 55545
Dutton Forshaw
Bircholt Rd, Parkwood, Maidstone (DJ BL LR)
☎(0622) 65461
Granville
Granville Rd, Maidstone
☎(0622) 61571
Haynes Bros
23 Ashford Rd, Maidstone
(Frd) ☎(0622) 56781
C A Marsh & Sons
57 Hartnup St, Maidstone
☎(0622) 27281
Rawden Rd
(MAK Eng) Rawden Rd, Maidstone ☎(0622) 678117
Rootes
Mill St, Maidstone (Tal Peu)
☎(0622) 53333
Tudor
(Dutton Forshaw) London Rd, Maidstone (DJ BL MG)
☎(0622) 53969

M20 Swanley – Hollingbourne 20 miles

From the edge of Greater London to the far side of Maidstone, this motorway might be the first leg of a journey to cross the Channel.

Opened in 1963, the Medway Bridge carries the M2 across the river near Rochester.

Junction 7

Another access to Maidstone here, with the road north crossing the Downs towards the Isle of Sheppey.

GARAGES
Highland
(P Heard Services) Detling
☎(0622) 39864
Westwood Motor Co
Thurnham Ln, Bearsted
☎(0622) 37333

Junction 8 R

The 'end' of the motorway, although it does continue at Ashford after 13 miles on the A20 – see next page for details. Eyehorne Manor lies a short distance north of this junction while a similar distance south brings you to Leeds Castle – the 'loveliest castle in the world'.

LOCAL RADIO STATIONS	Medium Wave		VHF/FM
	Metres	kHz	MHz
BBC Radio London	206	1458	94.9
BBC Radio Kent	290	1035	96.7
IBA Northdown Radio	314	954	NA

M2 Gillingham–Faversham   15 miles

The motorway continues through the Kent countryside towards Canterbury and the North Kent Coast.

The first Franciscan settlement in England was Greyfriars at Canterbury. Only this fine old building near the river survives.

Farthing Corner Services
(Rank) ☎(0634) 3343/4
Restaurant. Fast food
(summer). HGV Cafe.
Picnic area. Shop. Petrol.
Diesel. Breakdowns.
Repairs. HGV parking.

Long-term/overnight
parking for caravans £3
(max 24hrs). Baby-
changing. For disabled:
toilets, ramp **NB** steps in
building. Foot-bridge.
Credit cards accepted.

Junction 5

The famous Kent orchards are all around this stretch of motorway. At Sittingbourne is the Dolphin Yard Sailing Barge Museum where many old barges are being restored.

WHERE TO STAY
★★**Coniston**
London Rd, Sittingbourne
☎(0795) 23927
GH **Hillcroft Boarding House**
94 London Rd, Sittingbourne
☎(0795) 71501

GARAGES
Chalkwell
195 Chalkwell Rd,
Sittingbourne ☎(0795) 23982

Swale Motors
Crown Quay Ln,
Sittingbourne (Frd)
☎(0795) 70711
Upchurch
2 Horsham Ln, Upchurch,
Sittingbourne ☎(0634) 31684
E G Pritchard
London Rd, Bapchild (Vau
Opl Ren) ☎(0795) 76222

Junction 6

Faversham is an attractive and historic town on a navigable creek which flows into the Swale. The Fleur de lis Heritage Centre is housed in a 15th-century building and the Chart Gunpowder Mills, of 18th-century origin, have been restored.

WHERE TO EAT
❀××**Read's**
Painters Forstal, Faversham
☎(0795) 535344
GARAGES
East St
41 East St, Faversham
☎(0795) 532351
Forbes Rd
(Reeves & Stratford) Forbes
Rd, Faversham (Tal)
☎(0795) 536789
Invicta Motors
West St, Faversham (Frd)
☎(0795) 532255
Ospringe Rd
(G & E Newbery & Sons)
Ospringe Rd, Faversham
☎(0795) 534566

Junction 7

The motorway stops just six miles short of Canterbury, amidst yet more fruit orchards. Chilham Castle, to the south is the venue for a number of special events in its extensive grounds.

GARAGES
Woods
49 The Street, Boughton
☎(0227) 751307

M20 Ashford–Folkestone   14 miles

Between the Downs and Romney Marsh the motorway heads for the Channel Ports for crossings to Calais, Boulogne, Ostend, Zeebrugge and Dunkirk.

Junction 9

The market town of Ashford is the home of the Intelligence Corps Museum with items from the first World War to the present day. Just outside the town Godinton Park is a 17th century gabled house with a topiary garden.

WHERE TO STAY
★★★**Spearpoint**
Canterbury Rd, Kennington,
Ashford (EXEC)
☎(0233) 36863
★★★★⚑**Eastwell Manor**
Eastwell Park (3m N A251),
Ashford (PRE)
☎(0233) 35751
GH **Croft**
Canterbury Rd, Kennington,
Ashford ☎(0233) 22140

WHERE TO EAT
PS **Hothfield Common
Transit Picnic Site**
on A20 4m NW of Ashford
OS189 TQ9746.

GARAGES
Kennedys
Faversham Rd, Kennington,
Ashford (Vau Opl)
☎(0233) 23173
Beaver
18 Beaver Rd, Ashford
☎(0233) 35735

Junction 10

On the other side of Ashford, this junction is close to Swanton Mill at Mersham, a restored corn mill which won the 1975 European Architectural Heritage Year Award. At

Brook, to the north, is The Wye College Museum of Agriculture.

WHERE TO STAY
INN **New Flying Horse**
Upper Bridge St, Wye
☎(0233) 812297

WHERE TO EAT
××**Wife of Bath**
Wye ☎(0233) 812540

GARAGES
Taylors
150 Bridge St, Wye (BL)
☎(0233) 812331

Junction 11

South is the seaside town of Hythe, from where the Romney, Hythe and Dymchurch Railway follows the coast to Dungeness. At Lympne is a restored Norman to 15th-century castle and the famous Port Lympne Zoo Park.

WHERE TO STAY
☆☆**Royal Oak Motor Motel**
Ashford Rd, Newingreen
☎(0303) 66580
★★★★**Imperial**
Princes Pde, Hythe
☎(0303) 67441
★★★**Stade Court**
West Pde, Hythe (BW)
☎(0303) 68263
★**Swan**
59 High St, Hythe
☎(0303) 66236
GH **Dolphin Lodge**
16 Marine Pde, Hythe
☎(0303) 69565

WHERE TO EAT
Butt of Sherry
Theatre St, Hythe
☎(0303) 66112

GARAGES
Norringtons
Main Rd, Sellindge
☎(030381) 2120
Sellindge Service Station
Sellindge (Rel FSO)
☎(030381) 2181

Junctions 12 and 13

Folkestone is both a popular holiday resort and a busy Channel Port with regular services to France and Belgium.

WHERE TO STAY
★★**Burlington**
Earls Av, Folkestone
☎(0303) 55301
★★**Chilworth Court**
39–41 Earls Av, Folkestone
(MINO) ☎(0303) 41583
★★★**Clifton**
The Leas, Folkestone (THF)
☎(0303) 41231
GH **Belmonte**
30 Castle Hill Ave, Folkestone
☎(0303) 54470
GH **Arundel**
The Leas, 3 Clifton Rd,
Folkestone ☎(0303) 52442
GH **Argos**
6 Marine Ter, Folkestone
☎(0303) 54309
GH **Beaumont**
5 Marine Ter, Folkestone
☎(0303) 52740
GH **Horseshoe**
29 Westbourne Gdns,
Folkestone ☎(0303) 52184
GH **Wearbay**
25 Wearby Cres, Folkestone
☎(0303) 52586
FH Mrs J A Matthew **Lower
Arpinge Farm**, Arpinge
☎(0303) 78102

WHERE TO EAT
✕ **Paul's**
2a Bouvene St West,
Folkestone ☎(0303) 59697
✕ **Emilio's Portofino**
124a Sandgate Rd,
Folkestone ☎(0303) 55866
✕✕ **La Tavernetta**
Leaside Court, Folkestone
☎(0303) 54955
Pullman Wine Bar
7 Church St, Folkestone
☎(0303) 52524

GARAGES
Martin Walter
Caesar's Way, Cheriton,
Folkestone (Vau Opl Fia Cit)
☎(0303) 76431
Alsted
Unit 5, Ross Way, Shorncliffe
Trading Est, Folkestone (Peu
Tal) ☎(0303) 39639
Peacocks
Foord Rd North, Folkestone
(Frd) ☎(0303) 41234

The three-star Spearpoint Hotel is surrounded by five acres of lovely wooded parkland near Ashford. There are a large number of places of interest in this part of Kent and the orchard tour is lovely in spring.

Troy
75–79 Canterbury Rd,
Folkestone (Maz)
☎(0303) 42699
Hawkinge Auto Services
88 Canterbury Rd, Hawkinge
(Sko) ☎(030389) 2616

Densole Filling Station
(W J Montgomerie) 400
Canterbury Rd, Densole
☎(030389) 2379
Capel
30 Old Dover Rd, Capel le
Ferne ☎(0303) 54282

LOCAL RADIO STATIONS

	Medium Wave		VHF/FM
	Metres	kHz	MHz
BBC Radio Kent	290	1035	96.7
IBA Network East Kent	497	603	97.0
(due to open late '84 or early '85)			

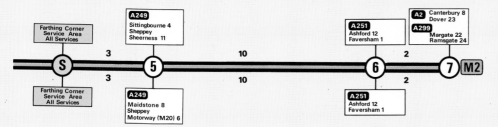

A249 Sittingbourne 4 Sheppey Sheerness 11

A251 Ashford 12 Faversham 1

A2 Canterbury 8 Dover 23
A299 Margate 22 Ramsgate 24

Farthing Corner Service Area All Services — **S** — 3 — **5** — 10 — **6** — 2 — **7** — M2

Farthing Corner Service Area All Services — 3 — 10 — 2

A249 Maidstone 8 Sheppey Motorway (M20) 6

A251 Ashford 12 Faversham 1

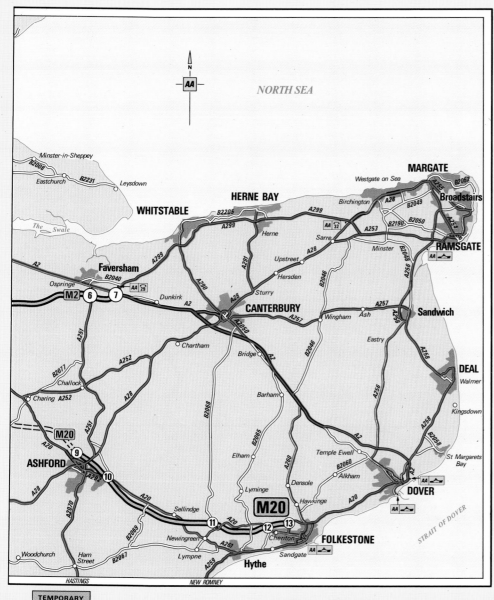

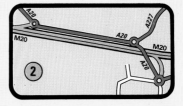

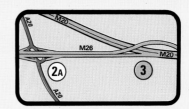

TEMPORARY MOTORWAY TERMINAL ACCESS TO EASTBOUND LANE OF A20

A292 Ashford 2
A20 Sellindge 5

B2068 Canterbury 13 Hythe (A261) 3

A20 Cheriton ½ Sandgate (A259) 2

A20 Dover 8 Folkestone & Harbour 2

4 — **8** — *projected* — **9** — 2 — **10** — 7 — **11** — 4 — **12** — 1 — **13** — M20

4 — — 2 — 7 — 4 — 1

ACCESS TO M20 FROM WESTBOUND LANE OF A20

PROJECTED

A20 ALL TRAFFIC

A292 Ashford 2

B2068 Lympne (A20) 2 Canterbury 13

A20 Cheriton ½ Lyminge (B2065) 4

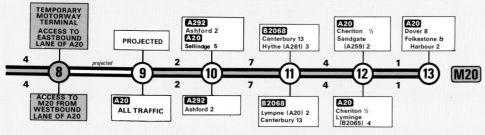

Odiham - Basingstoke M3

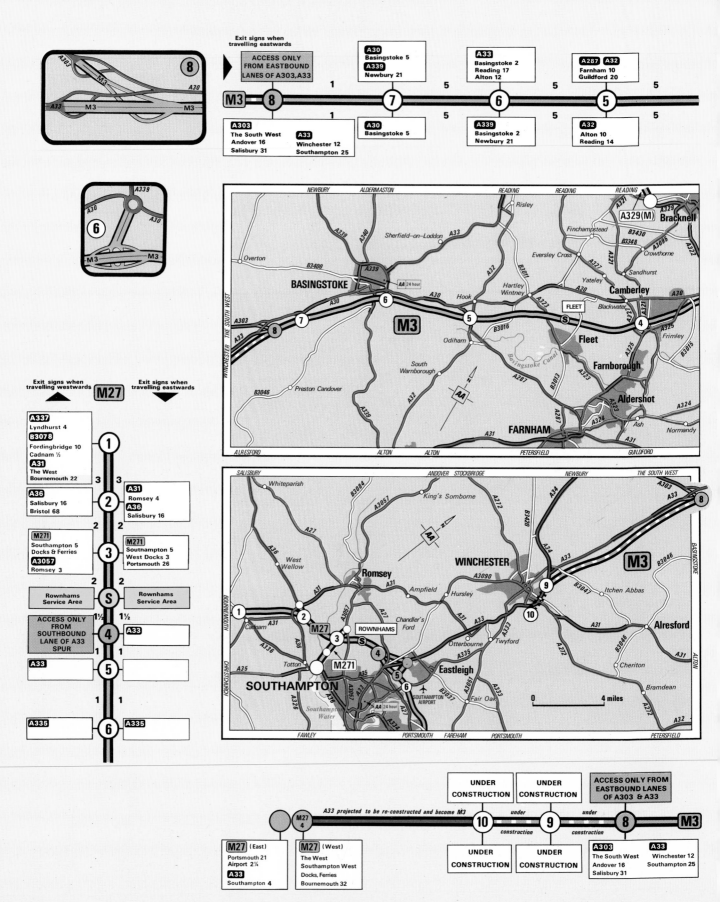

8

Exit signs when travelling eastwards

ACCESS ONLY FROM EASTBOUND LANES OF A303, A33

A30	A33	A287 A32
Basingstoke 5	Basingstoke 2	Farnham 10
A339	Reading 17	Guildford 20
Newbury 21	Alton 12	

M3 — 8 — 1 — 7 — 5 — 6 — 5 — 5 — 5

A303
The South West
Andover 16
Salisbury 31

A33
Winchester 12
Southampton 25

A30
Basingstoke 5

A339
Basingstoke 2
Newbury 21

A32
Alton 10
Reading 14

6

Exit signs when travelling westwards **M27** Exit signs when travelling eastwards

A337
Lyndhurst 4
B3078
Fordingbridge 10
Cadnam ½
A31
The West
Bournemouth 22

1

A31
Romsey 4
A36
Salisbury 16

A36
Salisbury 16
Bristol 68

2

M271
Southampton 5
Docks & Ferries
A3057
Romsey 3

3

M271
Southampton 5
West Docks 3
Portsmouth 26

Rownhams
Service Area

S

Rownhams
Service Area

ACCESS ONLY FROM SOUTHBOUND LANE OF A33 SPUR

4

A33

A33

5

A335

6

A335

UNDER CONSTRUCTION   UNDER CONSTRUCTION

ACCESS ONLY FROM EASTBOUND LANES OF A303 & A33

M27 4

A33 projected to be re-constructed and become M3

10 — under construction — 9 — under construction — 8 — M3

M27 (East)
Portsmouth 21
Airport 2¼
A33
Southampton 4

M27 (West)
The West
Southampton West
Docks, Ferries
Bournemouth 32

UNDER CONSTRUCTION   UNDER CONSTRUCTION

A303
The South West
Andover 16
Salisbury 31

A33
Winchester 12
Southampton 25

Junction 5

The navigable part of the Basingstoke Canal runs to the south of this junction through the pretty village of North Warnborough, then on past Odiham. This small town has a particularly attractive main street and is the home of an RAF station. To the north is Stratfield Saye, a large estate which was given to the Duke of Wellington by a grateful nation after his victory at Waterloo. The house is open to the public and part of the park has been developed as the Wellington Country Park.

WHERE TO STAY
☆☆**Raven**
Station Rd, Hook
☎(025672) 2541
GH **Oaklea**
London Rd, Hook
☎(025672) 2673

WHERE TO EAT
White Hart
London Rd, Hook
☎(025672) 2462
Whyte Lyon
Hartley Wintney
☎(025126) 2037

GARAGES
Odiham Motor Co
Hook Rd, North Warnborough
(Maz) ☎(025671) 2537
Clover Leaf
The Square, Odiham (MB)
☎(025671) 2294
Hampshire
Dunleys Hill, Odiham (BMW)
☎(025671) 2556

Junction 6

New development has all but swamped the market town of Basingstoke, although some parts of the old town centre survive. Industrial estates surround the town and there is an extensive

business area housing many national companies' headquarters. A short distance to the east of the town are the ruins of Basing House, a famous Civil War stronghold. To the north is The Vyne, an early 16th-century house now owned by the National Trust. Gliding enthusiasts will turn southwards here towards Lasham, on the Alton road, a well-known airfield for this sport.

WHERE TO STAY
☆☆☆**Crest**
Grove Rd, Basingstoke
(CRH) ☎(0256) 68181
★★**Red Lion**
London St, Basingstoke
(AHT) ☎(0256) 28525
GH **Wessex House**
Sherfield on Loddon
☎(0256) 882243

WHERE TO EAT
Corks Food and Wine Bar
25 London St, Basingstoke
☎(0256) 52622
Tundoor Mahal Restaurant
4 Winchester St, Basingstoke
☎(0256) 3795

GARAGES
Clover Leaf Cars
London Rd, Old Basing (Fia
Lnc AR) ☎(0256) 55221
Clover Leaf Cars
(Riverdene)
London Rd, Basingstoke
(Vlo) ☎(0256) 66111
J Davy
West Ham, Basingstoke (Vau
Opl) ☎(0256) 62551
Hadley
Houndmills, Basingstoke
(TAL Peu) ☎(0256) 65991
Jacksons
Lower Wote St, Basingstoke
(Frd) ☎(0256) 3561
Martins
The Hatch, London Rd,
Basingstoke (VW Aud)
☎(0256) 24444
Ralph Motors
Viables Ln, Basingstoke
☎(0256) 64204
W W Webber
New Loop Rd, Basingstoke
(BL DJ LR) ☎(0256) 24561

Wheeler & Ayland
Eastrop Rbt, Basingstoke
(Dai) ☎(0256) 65454

Junction 7

Beyond Basingstoke the motorway passes through hilly pastures. The pretty village of Dummer is just a short distance from this junction. To the north Overton is attractive, a village on the upper reaches of the River Test – one of our foremost trout rivers.

WHERE TO STAY
★★**Beach Arms**
(on B3400) Oakley
☎(0256) 780210

GARAGES
Beach Arms Service Station
Oakley ☎(0256) 780331
Wheatsheaf
North Waltham
☎(025675) 254

Junction 8 R

For the time being the motorway ends here at Popham on the edge of the Micheldever Forest. Another section is under construction and when complete (sometime during 1985) the motorway will extend to Winchester.

WINCHESTER

Although it is not yet on the motorway route, it is likely that many travellers who leave the M3 at its end will be continuing in this direction. Winchester would certainly be a pleasant and interesting place in which to break your journey.

Surrounded by beautiful countryside, Winchester is an ancient city which has retained a great deal of character. The High Street is watched over from its eastern end by a statue of King Alfred who held court here. Winchester was, in fact, the capital of England until the Norman Conquest and its 900-year-old cathedral contains the mortuary chests of a number of Anglo Saxon kings and bishops. King Cnut and his wife are also entombed there. A particularly interesting memorial in the cathedral is that of William Walker, the deep sea diver who went down into the peat bog beneath the building to replace the medieval timbers with concrete, so

saving this fine building from sinking and eventually crumbling.

The area around the cathedral contains many lovely old buildings including those of Winchester College, one of Britain's leading public schools which was founded in 1382. The Pilgrims' Hall in the Inner Close also dates from the 14th century and forms part of Pilgrims' School.

A delightful riverside walk follows the River Itchen, skirting the old wall of Wolvesey Castle, built as a bishop's palace in the 12th century and destroyed by Cromwell. The new Bishop's Palace of 1684 and some remains of a Roman wall can also be seen.

At the other end of the town is the Great Hall, famous for containing 'King Arthur's Round Table'. This massive circle of oak measuring 18ft across and weighing 1¼ tons is known to be more than 500 years old, but it is almost certain that Arthur and his knights, whose names decorate its border, never sat at it.

The ruin of Odiham Castle, from where King John set out to sign Magna Carta in 1215 stands on the outskirts of the town near Basingstoke Canal at North Warnborough.

M3 Odiham–Basingstoke   11 miles

The first part of this motorway is dealt with on the following page. Here we proceed through the Hampshire countryside, past the growing commercial centre of Basingstoke and on towards the lovely old city of Winchester.

Basingstoke's Red Lion is a 17th-century two-star inn at the centre of the old upper town.

WHERE TO STAY
★★★ ⬚B L **Lainston House**
Sparsholt ☎(0962) 63588
★★★★**Wessex**
Paternoster Row, Winchester
(THF) ☎(0962) 61611
★★**Westacre**
Sleepers Hill, Winchester
☎(0962) 68403

WHERE TO EAT
The Cart and Horses Inn
Kings Worthy
☎(0962) 882360
Mr Pitkin's Wine Bar and Eating House
4 Jewry St, Winchester
☎(0962) 696630
✕✕**Splinters**
9 Gt Minster St, Winchester
☎(0962) 64004
Minstrels Restaurant
18 Little Minster St,
Winchester ☎(0962) 67212
Old Mill Restaurant
1 Bridge St, Winchester
☎(0962) 63151

The Old Chesil Rectory
Chesil St, Winchester
☎(0962) 53177
PS **Avington Park**
1m S of Itchen Abbas
OS185 SU5332
PS **Crab Wood**
2m W of Winchester off A3090
OS185 SU4329
PS **Farley Mount**
W of Winchester on
unclassified road
OS185 SU4229

GARAGES
Easton
(Mould & Thompson) Easton,
Winchester ☎(096278) 319
Curtis Motors
London Rd, Kings Worthy
☎(0962) 880269
Springvale Co
Kings Worthy
☎(0962) 880803

LOCAL RADIO STATIONS

	Medium Wave		VHF/FM
	Metres	kHz	MHz
BBC Radio Oxford	202	1485	95.2
BBC Radio Solent	300	999	96.1
IBA Radio 210	210	1431	97.0
IBA County Sound	203	1476	96.6

M27 Cadnam–Eastleigh

A Thermo-chart is included opposite to provide a quick reference to the junctions on this motorway, but full details can be found on pages 78–79.

M3 London–Fleet 22 miles

The motorway begins at Sunbury and heads westwards through well-wooded Surrey countryside.

Junction 1

Sunbury lies in a loop of the River Thames amidst a chain of large reservoirs. Close to this junction is Kempton Park racecourse.

WHERE TO STAY
☆☆☆**Shepperton Moat House**
Felix Ln, Shepperton (QM)
☎(0932) 241404

WHERE TO EAT
××**Terrazza**
45 Church St, Ashford
☎(07842) 44887
×**La Bussola**
32a High St, Walton-on-Thames ☎(0932) 244889
××**Lantern**
20 Bridge Rd, East Molesey
☎01–979 1531

GARAGES
Sunbury Service Station
115 Staines Rd West,
Sunbury-on-Thames
☎01–572 9678
D M Autos
Ashford Ind Est, Shield Rd,
Ashford ☎01–572 9678
Spring Grove Car & Commercial Maintenance
Browells Ln, Feltham
☎01–751 5366
Ashmans
284–286 Kingston Rd,
Ashford (Vau)
☎(0784) 52763

Clock House Service Station
Clockhouse Ln, Bedfont
☎(0784) 41731
Fern Automobiles
Hounslow Rd, Feltham
☎01–890 0395

The Thames at Runnymede near Egham, famous for the signing of Magna Carta.

Broad Lane
163 Broad Ln, Hampton
☎01–979 2233

Walchry Motors
143 Hersham Rd, Walton-on-Thames (Tal Peu)
☎(0932) 223761

Junction 2

The Motorway connects here with the M25 London Orbital Motorway – see pages 74–77 for details. Alongside the motorway is Thorpe Park, a 500 acres theme park, although access from the motorway is circuitous.

WHERE TO STAY
★★**Pack Horse**
Thames St, Staines (AHT)
☎(0784) 54221
☆☆☆**Runnymede**
Windsor Rd, Egham
☎(0784) 36171

GARAGES
Ayebridges
(MGM Motor Co) Thorpe Lea Rd, Egham ☎(0784) 56100
Motest
Egham Hill, Egham
☎(0753) 42666
Henlys
Unit 23B, Central Trading Est, Staines (BL) ☎(0784) 55281
G Blair Addlestone
Chertsey Rd, Addlestone (Tal Peu) ☎(0932) 42242
Trident
Guildford Rd, Ottershaw (BL LR) ☎(093287) 2561

Junction 3

Ascot racecourse is to the north, famous for its fashionable Royal Ascot meeting in June.

WHERE TO STAY
☆☆☆☆**Pennyhill Park**
College Ride, Bagshot (PRE)
☎(0276) 71774
★★**Cricketers**
(Henekey's) London Rd,
Bagshot (THF)
☎(0276) 73196
☆☆☆**Berystede**
Bagshot Rd, Sunninghill
(THF) ☎(0990) 23311
GH **Highclere House**
Kings Rd, Sunninghill, Ascot
☎(0990) 25220

WHERE TO EAT
××**Sultans Pleasure**
13 London Rd, Bagshot
☎(0276) 75114
PS **Bracknell**
Bagshot Road, Transit Picnic Site OS175 SU9065

GARAGES
Brandon Service Station
Guildford Rd, Lightwater
(Ren) ☎(0276) 73060
Woodville Motors
Guildford Rd, Lightwater
☎(0276) 76273
Fina Filling Station
(D J Warrington Co) London
Rd, Bagshot ☎(0276) 71696
PIRA Autos
40 High St, Sunningdale (Col)
☎(0990) 24866

Junction 4

A great military concentration surrounds this junction – the aircraft establishment at Farnborough; the army at Aldershot, Church Crookham and, of course, Sandhurst. Visit the Airborne Forces Museum and the Royal Corps of Transport Museum in Aldershot, the Gurkha Museum at Church Crookham and the RAMC Historical Museum at Ash Vale.

WHERE TO STAY
★★★**Frimley Hall**
Portsmouth Rd, Camberley
(THF) ☎(0276) 28321
☆☆☆**Queens**
Lynchford Rd, Farnborough
(AHT) ☎(0252) 545051
☆☆☆**Waterloo**
Dukes Ride, Crowthorne
(AHT) ☎(0344) 777711

GARAGES
P Warren & Co
Frimley Green (Ren)
☎(0252) 835436
Chilton Service Station
(A & R Motors) Fleet Rd, Cove
☎(0252) 545857
Berdan
144 High St, Sandhurst (Vlo)
☎(0252) 877333
Deepcut
88 Deepcut Bridge Rd,
Deepcut (BL) ☎(02516) 5631

Fleet Services
(THF) ☎(0984) 21656
Cafeteria. Waitress-service 'Little Chef' (west side). Shop. Petrol. Diesel. Breakdowns. Repairs. HGV parking. Long-term/overnight parking for caravans £2. Baby-changing. For disabled: toilets, ramp. Footbridge. Credit cards – shop, garage, restaurants.

M23 Hooley–Pease Pottage 16 miles

Well used by travellers to and from Gatwick Airport, this motorway will also set you in the right direction for the south coast at Brighton. The stretch from Junction 1 to Junction 7 is unlikely to be built in the foreseeable future.

Junction 7 R

The motorway begins just north of Reigate, a pleasant, if somewhat congested, town beneath the downs. The Priory, founded in 1235, but converted in Tudor times, is now a school with a small museum open to the public.

WHERE TO STAY
★★**Reigate Manor**
Reigate Hill, Reigate (BW)
☎(07372) 40125
☆☆**Pickard Motor**
Brighton Rd, Burgh Heath
(BW) ☎(07373) 57222
GH **Ashleigh House**
39 Redstone Hill, Redhill
☎(0737) 64763
GH **Priors Mead**
Blanford Road, Reigate
☎(07372) 48776
GH **Cranleigh**
41 West St, Reigate (Minotel)
☎(07372) 40600

WHERE TO EAT
Crocks
33 High St, Redhill
☎(0737) 61177
Cobbles Wine Bar
Brewery Yard, Bell St, Reigate
☎(07372) 44833

Home Maid
10 Church St, Reigate
☎(07372) 48806
New Hong Kong
27 Bell St, Reigate
☎(07372) 43374
PS **Farthing Downs**
1m S of junction A23/B2030
OS187 TQ3057
PS **Coulsdon Common**
1½m NW of Caterham

GARAGES
Oliver Taylor Recovery
Coach Station, Banstead Rd,
Caterham ☎(0883) 42341
Denyer Motors
204 Godstone Rd, Caterham
☎(0883) 42240
French & Foxwell
Brighton Rd, Burgh Heath
(Vau Opl) ☎(07373) 53366

The three-star Crest Hotel stands near Gatwick.

Taylor Autos
Reading Arch Rd, Redhill
☎(0757) 62884

Junction 8

This junction gives access to and from the M25 London Orbital motorway – see pages 74–77 for details.

Junction 9

This is the Gatwick Airport turning. A little farther west, at Charlwood, flyers of a different kind can be seen at Gatwick Zoo and Aviaries.

WHERE TO STAY
☆☆☆**Gatwick Hilton International**
Gatwick Airport
☎(0293) 518080
☆☆**Chequers Thistle**
Brighton Rd, Horley (TS)
☎(02934) 6992
☆☆**Gatwick Moat House**
Longbridge Rbt, Horley (QM)
☎(02934) 5599
☆☆**Gatwick Penta**
Povey Cross Rd, Horley
☎(02934) 5533
☆☆**Post House**
Povey Cross Rd, Horley (THF)
☎(02934) 71621

GARAGES
Horley Motor Co
61 Brighton Rd, Horley (Ren)
☎(02934) 72566
Auto Services
Massetts Rd, Horley (BL)
☎(02934) 5176

Junction 10

Crawley, accessible from this junction, is a new town built around an old one.

WHERE TO STAY
☆☆☆**Copthorne**
Copthorne Rd, Copthorne
☎(0342) 714971
☆☆☆**Gatwick Manor**
London Rd, Crawley (BL)
☎(0293) 26301
☆☆☆**Gatwick Concorde**
Lowfield Heath, Crawley
(QM) ☎(0293) 33441
☆☆**Crest**
Tushmore Rbt, Langley Dr,
Crawley (CRH)
☎(0293) 29991
GH **Barnwood**
Balcombe Rd, Pound Hill,
Crawley ☎(0293) 882709
GH **Gatwick Skylodge**
London Rd, County Oak,
Crawley ☎(0293) 514341

WHERE TO EAT
×**Fox and Hounds**
Crawley ☎(096272) 285

GARAGES
Crawley Down
Snowhill, Copthorne (BL)
☎(0342) 713933
Southern Counties
263–269 Haslett Av, Three
Bridges, Crawley (BL)
☎(0293) 27101
A J Still
(Motors) Norman House,
Stephenson Way, Three
Bridges, Crawley
☎(0293) 27067
Central
(Fields) 163 Three Bridges
Rd, Crawley (Cit)
☎(0293) 25533
Evans Halshaw Sussex
Fleming Way, Crawley (Vau)
☎(0293) 29771

HRS Recoveries
Dales Coach Yard, Spindle
Way, Crawley ☎(0293) 26357

Junction 11

The motorway ends south of Crawley on the edge of St Leonards forest. Lovely gardens to visit include Leonardslee Gardens at Lower Beeding and Nymans at Handcross.

WHERE TO STAY
★★★**Goffs Park**
45 Goffs Park Rd, Crawley
☎(0293) 35445
★★★**George**
High St, Crawley (THF)
☎(0293) 24215

WHERE TO EAT
Solomons Ancient Priors Restaurant and Wine Bar,
High St, Crawley
☎(0293) 36223

GARAGES
C Gadsdon
5 Brighton Rd, Southgate,
Crawley (Dat) ☎(0293) 35264
Southern Counties
Ifield Road, Crawley (BL DJ
LR Rar) ☎(0293) 20191

LOCAL RADIO STATIONS

	Medium Wave		VHF/FM
	Metres	*kHz*	*MHz*
BBC Radio London	206	1458	94.9
IBA Capital Radio	194	1548	95.8
IBA LBC	261	1152	97.3
IBA County Sound	203	1476	96.6

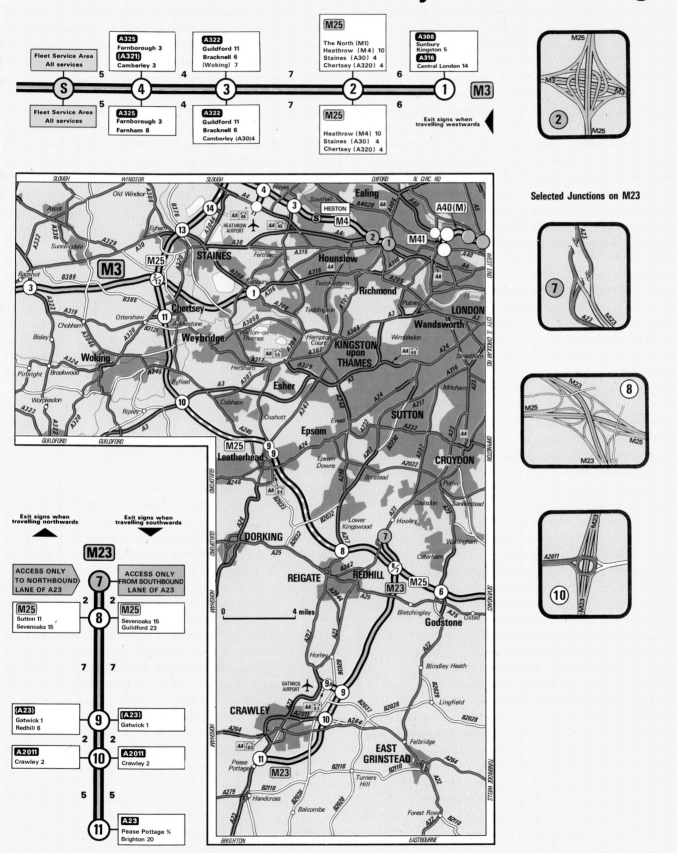

M25
The North (M1)
Heathrow (M4) 10
Staines (A30) 4
Chertsey (A320) 4

A308
Sunbury
Kingston 5
A316
Central London 14

A325
Farnborough 3
(A321)
Camberley 3

A322
Guildford 11
Bracknell 6
(Woking) 7

Fleet Service Area
All services

S   **4**   **3**   **2**   **1**   **M3**

5   4   7   6

5   4   7   6

Fleet Service Area
All services

A325
Farnborough 3
Farnham 8

A322
Guildford 11
Bracknell 6
Camberley (A30) 4

M25
Heathrow (M4) 10
Staines (A30) 4
Chertsey (A320) 4

Exit signs when
travelling westwards

M25

2

M3   **M3**

M25

Selected Junctions on M23

7

8

10

Exit signs when
travelling northwards

Exit signs when
travelling southwards

M23

ACCESS ONLY
TO NORTHBOUND
LANE OF A23

7

ACCESS ONLY
FROM SOUTHBOUND
LANE OF A23

M25
Sutton 11
Sevenoaks 15

2   2

8

M25
Sevenoaks 15
Guildford 23

7   7

(A23)
Gatwick 1
Redhill 6

9

(A23)
Gatwick 1

2   2

A2011
Crawley 2

10

A2011
Crawley 2

5   5

11

A23
Pease Pottage ¾
Brighton 20

London - Theale M4

Bracknell - Reading A329(M)

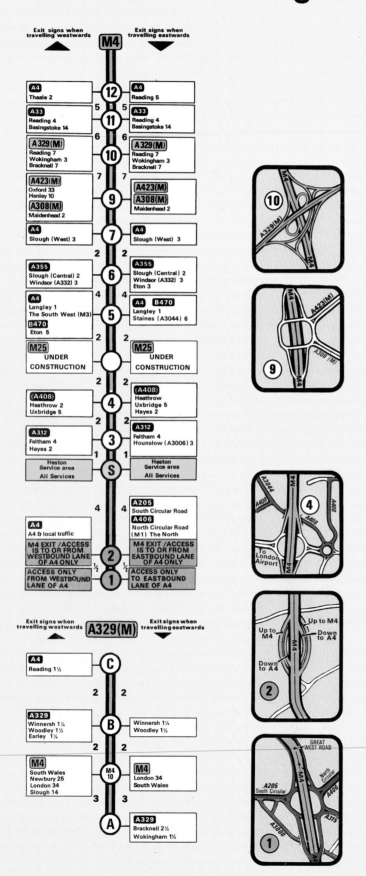

Exit signs when travelling westwards ▲ M4 Exit signs when travelling eastwards ▼

A4
Theale 2 | 12 | **A4**
Reading 5

5 | | 5

A33
Reading 4
Basingstoke 14 | 11 | **A33**
Reading 4
Basingstoke 14

6 | | 6

A 329(M)
Reading 7
Wokingham 3
Bracknell 7 | 10 | **A329(M)**
Reading 7
Wokingham 3
Bracknell 7

7 | | 7

A423(M)
Oxford 33
Henley 10

A308(M)
Maidenhead 2 | 9 | **A423(M)**
A308(M)
Maidenhead 2

A4
Slough (West) 3 | 7 | **A4**
Slough (West) 3

2 | | 2

A355
Slough (Central) 2
Windsor (A332) 3 | 6 | **A355**
Slough (Central) 2
Windsor (A332) 3
Eton 3

4 | | 4

A4
Langley 1
The South West (M3)

B470
Eton 5 | 5 | **A4** **B470**
Langley 1
Staines (A3044) 6

2 | | 2

M25
UNDER
CONSTRUCTION | ○ | **M25**
UNDER
CONSTRUCTION

2 | | 2

(A408)
Heathrow 2
Uxbridge 5 | 4 | **(A408)**
Heathrow
Uxbridge 5
Hayes 2

2 | | 2

A312
Feltham 4
Hayes 2 | 3 | **A312**
Feltham 4
Hounslow (A3006) 3

1 | | 1

Heston
Service area
All Services | S | Heston
Service area
All Services

| | 4 | **A205**
South Circular Road
A406
North Circular Road
(M1) The North

4 | | 4

A4
A4 & local traffic

**M4 EXIT /ACCESS
IS TO OR FROM
WESTBOUND LANE
OF A4 ONLY** | 2 | **M4 EXIT /ACCESS
IS TO OR FROM
EASTBOUND LANE
OF A4 ONLY**

**ACCESS ONLY
FROM WESTBOUND
LANE OF A4** | 1 | ½ **ACCESS ONLY
TO EASTBOUND
LANE OF A4**

Exit signs when travelling westwards ▲ A329(M) Exit signs when travelling eastwards ▼

A4
Reading 1½ | C

2 | | 2

A329
Winnersh 1¼
Woodley 1½
Earley 1¼ | B | Winnersh 1¼
Woodley 1½

2 | | 2

M4
South Wales
Newbury 25
London 34
Slough 14 | M4
10 | **M4**
London 34
South Wales

3 | | 3

| A | **A329**
Bracknell 2½
Wokingham 1½

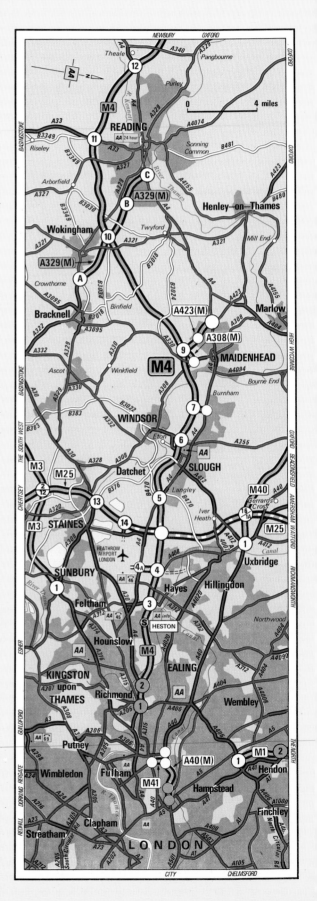

M4 London – Theale   38½ miles

This busy stretch of motorway takes traffic due west out of London through almost continuously built-up areas until it emerges into the Berkshire countryside. Heathrow Airport is en route.

Junctions 1 and 2 R

To the south the Royal Botanic Gardens, Old Deer Park and Syon Park form a lovely green area beside the Thames.

WHERE TO STAY
GH **Chiswick**
73 Chiswick High Rd, Chiswick ☎01–994 1712
☆☆☆**Carnarvon**
Ealing Common
☎01–992 5399
GH **Grange Lodge**
50 Grange Rd, Ealing
☎01–567 1049
☆☆☆**Cunard International**
1 Shortlands, Hammersmith
☎01–741 1555

WHERE TO EAT
Fouberts Wine Bar
162 Chiswick High Rd, Chiswick ☎01–994 5202
Maids of Honour
288 Kew Rd, Kew
☎01–940 2752
Le Provence
14 Station Pde, Kew
☎01–940 6777

GARAGES
Warwick Wright Motors
Chiswick Rbt (Peu Tal)
☎01–995 1466
Fountain Service Station
Hogarth Rbt, Gt West Rd, Chiswick ☎01–994 2446
Hogarth
120 Cranbrook Rd, Chiswick (BL) ☎01–994 1356
Vehicle Maintenance
Mill Hill Ter, Acton
☎01–992 7741
Chipstead Recovery
(Chipstead of Kensington)
6 Stamford Brook Rd, Hammersmith
☎01–572 9678

Heston Services

(Granada) ☎
01–574 7271
Restaurant. Fast food (south side). Shop. Petrol. Diesel. HGV parking. Long term/overnight parking for caravans £3.50. Baby-changing. For disabled: toilets, ramp. Credit cards – shop, garage, restaurant.

With a large international clientele, the Holiday Inn provides a luxurious welcome for travellers passing through Heathrow Airport.

Junction 3

The exit for Hillingdon and Hounslow.

WHERE TO STAY
☆☆**Berkeley Arms**
Bath Rd, Cranford, Heathrow Airport (EH) ☎01–897 2121
☆☆**Master Robert Motel**
366 Great West Rd, Hounslow, Heathrow Airport
☎01–570 6261

WHERE TO EAT
✕**Hounslow Chinese**
261–263 Bath Rd, Hounslow
☎01–570 2161
Travellers Friend
480 Bath Rd, Hounslow
☎01–897 8847

GARAGES
Rectory Autos
The Old Rectory, Church Rd, Cranford ☎01–897 0701
North Hyde Service Station
North Hyde Rd, Hayes
☎01–573 6912
Airways
242 Bath Rd, Hayes
☎01–759 9661

Junction 4

The spur to Heathrow Airport leaves the main motorway here.

WHERE TO STAY
☆☆☆☆**Holiday Inn**
Stockley Rd, West Drayton (HI) ☎(0895) 445555
☆☆☆**Post House**
Sipson Rd, West Drayton, Heathrow Airport (THF)
☎01–759 2323
☆☆☆**Excelsior**
Bath Rd, West Drayton, Heathrow Airport (THF)
☎01–759 6611
☆☆☆**Heathrow Penta**
Bath Rd, Hounslow, Heathrow Airport
☎01–897 6363
☆☆☆**Ariel**
Bath Rd, Hayes, Heathrow Airport (THF) ☎01–759 2552
★★★**Skyway**
Bath Rd, Hayes, Heathrow Airport (THF) ☎01–759 6311

☆☆☆☆**Sheraton Heathrow**
Colnbrook Bypass, West Drayton, Heathrow Airport
☎01–759 2424

GARAGES
Central
(Radley Autos) 1 High St, Harmondsworth
☎01–897 2051

Junction 5

To the north-west, Slough is a busy commercial centre.

WHERE TO STAY
☆☆☆**Holiday Inn**
Ditton Rd, Langley, (CHI)
☎(0753) 44244
GH **Francis House**
21 London Rd, Langley
☎(0753) 22286

WHERE TO EAT
✕**Chez Petit Laurent**
Country Life House, Slough Rd, Datchet ☎(0753) 49314

GARAGES
Golden Cross Service Station
(MOTEST) Old Bath Rd, Colnbrook ☎(0753) 42666
Friary Motors
Straight Rd, Old Windsor (Frd BL) ☎(07535) 61402

Junction 6

Southwards is Windsor with its splendid castle, a museum of the Household Cavalry and an exhibition depicting Queen Victoria's Diamond Jubilee celebrations. South-west of the town is Windsor Safari Park.

WHERE TO STAY
★★★**Castle**
High St, Windsor (THF)
☎(07535) 51011
★★★★**Oakley Court**
Windsor Rd, Water Oakley
☎(0628) 74141

WHERE TO EAT
✕✕**Don Peppino**
30 Thames St, Windsor
☎(07535) 60081

✕✕**Winsor's Peking**
14 High St, Windsor
☎(07535) 64942
Choices
10 Thames St, Windsor
☎(07535) 66437
Summerfield's
15–17 Thames St, Windsor
Drury House Restaurant
4 Church St, Windsor
☎(07535) 63734
PS **Alexandra Gardens**
River St or Barry Av, Windsor

GARAGES
Delta
(Delta Motor Co) 195–199 Clarence Rd, Windsor (Lad)
☎(07535) 60707
Stag Motors
2 Elm Rd, Windsor
☎(07535) 64143
Windrush
57 Farnham Rd, Slough (VW Aud) ☎(0753) 33917

Junction 7

To the north are Farnham Common and Burnham Beeches. Cliveden (NT), former home of the Astors, can be visited.

WHERE TO STAY
GH **Norfolk House**
Bath Rd, Taplow
☎(0628) 23687

WHERE TO EAT
PS **Burnham Beeches**
½m W of A355 at Farnham Common OS175 SU9585

GARAGES
White Heather
Village Rd, Dorney
☎(0628) 3188
Normans Conquest Service Station
372 Bath Rd, Cippenham, Slough (Dat Tal VW Aud)
☎(06286) 61797
Vehicle Repairs
361 Bath Rd, Slough
☎(06286) 63343
Station
Station Rd, Taplow (AR Col Lot Mgn BMW Kaw Rel)
☎(06286) 5353
Maidenhead Autos
Bath Rd, Taplow (Frd)
☎(0628) 29711

Junction 9

Maidenhead is a popular Thames-side town. Oldfield House is open to the public and, west of the town, is the Courage Shire Horse Centre.

WHERE TO STAY
☆☆☆☆**Crest**
Manor Ln, Maidenhead (CRH) ☎(0628) 23444
✿★★★**Fredrick's**
Shoppenhangers Rd, Maidenhead ☎(0628) 35934
★★**Bear**
High St, Maidenhead (AHT)
☎(0628) 25183

WHERE TO EAT
✕✕**Franco's La Riva**
Raymead Rd, Maidenhead
☎(0628) 33522
✕**Chef Peking**
74 King St, Maidenhead
☎(0628) 32851

✕**Chez Michel et Valerie**
Bridge Av, Maidenhead
☎(0628) 22450
✕**Maidenhead Chinese**
45–47 Queen St, Maidenhead ☎(0628) 24545
Bacchus Swiss Restaurant
St Mary's Walk, Maidenhead
☎(0628) 36638

GARAGES
D Ruskin
107 Windsor Rd, Bray (Dat)
☎(0628) 37535
Delta Motor Co
37–49 Grenfell Rd, Maidenhead (Lnc)
☎(0628) 22660
Lex Mead
128 Bridge Rd, Maidenhead
(BL DJ RR LR)
☎(0628) 33188

Junction 10

The A329(M) crosses here, giving access to the establishments listed below.

WHERE TO STAY
★★★**St Annes Manor**
London Rd, Wokingham
☎(0734) 784427
★★★**White Hart**
Sonning ☎(0734) 692277

GARAGES
Whitehouse Motors
(E J Barter) Reading Rd, Winnersh ☎(0734) 783718
Emmbrook Service Station
123–125 Reading Rd, Wokingham ☎(0734) 861860
Gowrings
Handpost Cnr, Finchampstead Rd, Wokingham (Frd)
☎(0734) 780873
Grove Service Station & Whitebridge Garage
Old Bath Rd, Twyford
☎(0734) 340402

Junction 11

The busy commercial town of Reading also has an interesting Museum of English Rural Life at the university.

WHERE TO STAY
☆☆☆**Post House**
Basingstoke Rd, Reading (THF) ☎(0734) 875485
★★**Ship**
Duke St, Reading (AHT)
☎(0734) 583455
O **Ramada**
Oxford Rd, Reading
☎(0734) 586222

GH **Private House**
98 Kendrick Rd, Reading
☎(0734) 874142

WHERE TO EAT
✿✕✕**Milton Sandford**
Church Rd Ln, Shinfield
☎(0734) 883783
✕✕**Mill House**
Basingstoke Rd, Swallowfield
☎(0734) 883124
Beadles Wine Bar
83 Broad St, Reading
☎(0734) 53162
Mama Mia
11 St Mary's Butts, Reading
☎(0734) 581357
Sweeney & Todd
10 Castle St, Reading
☎(0734) 585466

GARAGES
Hearn Bros
The Garage, Basingstoke Rd, Three Mile Cross
☎(0734) 883581
M Spence
Arborfield Rd, School Gn, Shinfield (Lnc Mgn Lot)
☎(0734) 883140
Horncastle
Bath Rd, Reading (Frd)
☎(0734) 412021
Penta
Penta House, Basingstoke Rd, Reading (BL DJ LR Rar)
☎(0734) 875151
D Ruskin
660 Wokingham Rd, Earley, Reading (Dat)
☎(0734) 669621
Whitley Wood
(Reading Toyota Ctr)
Basingstoke Rd, Reading
(Toy) ☎(0734) 871278
Zenith
160 Basingstoke Rd, Reading (Frd) ☎(0734) 875333
Williams
7 Weldale St, Reading
☎(0734) 53832

Junction 12

To the north, Pangbourne is a pleasant Thames-side village and Theale lies to the south west.

WHERE TO STAY
GH **Aeron**
191 Kentwood Hill, Tilehurst, Reading ☎(0734) 24119

WHERE TO EAT
✕✕**Knights Farm**
Burghfield ☎(0734) 52366

GARAGES
Theale Motor Works
22–24 High St, Theale (Ren)
☎(0734) 302422
Julians
38 Portman Rd, Reading (BL DJ LR Rar) ☎(0734) 585011

LOCAL RADIO STATIONS			
	Medium Wave		VHF/FM
	Metres	*kHz*	*MHz*
BBC Radio London	206	1458	94.9
BBC Radio Oxford	202	1485	95.2
IBA Capital Radio	194	1548	95.8
IBA LBC	261	1152	97.3
IBA Radio 210	210	1431	97.0

M4 Theale – Chippenham   50 miles

This is horse-racing country, passing Newbury with its racecourse and the Lambourne Downs where many trainers have their stables.

Junction 12 – see page 51

Junction 13

The old market town of Newbury is fast becoming the micro-chip capital of Britain, with many computer-related companies making it their home. New development has not spoilt the centre with its canal-side walks. The Old Cloth Hall, dating from the 17th century, is a fine half-timbered building housing the town museum.

WHERE TO STAY
★★★**Chequers**
Oxford St, Newbury (THF)
☎(0635) 43666
INN **Hare & Hounds**
Speen, Newbury
☎(0635) 47215

WHERE TO EAT
✕**La Riviera**
26 The Broadway, Newbury
☎(0635) 47499
✕**Sapient Pig**
29 Oxford Rd, Newbury
☎(0635) 47425
Cromwell's Wine Bar
20 London Rd, Newbury
☎(0635) 40255
Hatchet
Market Pl, Newbury
☎(0635) 47352

GARAGES
D White Motor Vehicle Repairs
Bates Yard, Curridge
☎(0635) 200236
Chieveley Motor Co
Long Ln, Hermitage (Jep Dai)
☎(0635) 200637
Gowrings
256 London Rd, Newbury
(Frd) ☎(0635) 45100
Nias
London Rd, Newbury (BL)
☎(0635) 41100
Wheelers
London Rd, Newbury (Ren)
☎(0635) 41020

Auto Centre
Ampere Rd, London Rd
Industrial Estate, Newbury
☎(0635) 44519
Edwards
Faraday Rd, London Rd
Industrial Estate, Newbury
☎(0635) 49624
Nias
172–174 Andover Rd,
Newbury (Dat)
☎(0635) 48515

Junction 14

Hungerford is a most attractive town which, before the arrival of the motorway, was on the main road west – the Bath Road. It is also an antique hunters paradise. A short distance to the west is Littlecote House, a lovely Tudor manor with a unique collection of Cromwellian armour. In the grounds a Roman temple has been excavated and there is also a replica wild west town.

WHERE TO STAY
★★**Elcot Park**
(1m N of A4) Elcot (BW)
☎(0488) 58100
★★★**Bear**
Hungerford (BW)
☎(0488) 82512
INN **Five Bells**
Wickham ☎(048838) 242

WHERE TO EAT
❀✕✕**Dundas Arms**
Kintbury ☎(0488) 58263
Tutti Pole
3 High St, Hungerford
☎(04886) 2515

GARAGES
P Stirland
17–19 Bridge St, Hungerford
(Frd) ☎(0488) 83678
Normans
Bath Rd, Hungerford (Peu Tal
Fia) ☎(0488) 82033

Membury Services
(Welcome Break)
☎(0488) 71880
24hr Cafeteria. Waitress-service restaurant. Hostess-service Buttery. Fast food. HGV cafe. Picnic area. Shop. Vending machines. Playground. Petrol. Diesel. Liquid Petroleum Gas. Repairs. HGV parking. Long-term/overnight parking for caravans £3. Baby-changing. For disabled: toilets, ramp. Footbridge.

Junction 15

Swindon is the industrial and commercial centre of Wiltshire and has seen much new development in recent years. The award-winning Brunel Shopping Centre covers some 13 acres with covered arcades. The town has a railway heritage and a fine Railway Museum is situated alongside the recently renovated GWR Railway Village where one house, restored to its turn-of-the-century condition, is open to the public.

WHERE TO STAY
☆☆☆**Post House**
Marlborough Rd, Swindon
(THF) ☎(0793) 24601
★★★**Goddard Arms**
High St, Old Town, Swindon
☎(0793) 692313
★★★**Wiltshire**
Fleming Way, Swindon (KH)
☎(0793) 28282
☆☆☆**Crest**
Oxford Rd, Stratton St
Margaret, Swindon (CRH)
☎(0793) 822921

WHERE TO EAT
Sheraton Suite
East St, Swindon
☎(0793) 24114

GARAGES
Greens
Marlborough Rd, Swindon
(Dat) ☎(0793) 27251
British Body Builders
Hooper Pl, Newport St,
Swindon ☎(0793) 27484
Skurrays
High St, Swindon (Vau Opl
Fia) ☎(0793) 20971
Swindon Autos
Drove Rd, Swindon (BL)
☎(0793) 34035
Dorcan Way Service Station
Dorcan Way, Swindon
☎(0793) 25449
Bath Road
Bath Rd, Swindon (Hon)
☎(0793) 24217
Green Meadow Service Station
Thames Av, Swindon
☎(0793) 35895
Martin Whale Engineering
Unit 25B, Techno Trading
Estate, Bramble Rd, Swindon
☎(0793) 37181

Junction 16

One mile north of this junction is Lydiard Park, a fine Georgian mansion set in pleasant parkland. The old town of Wootton Bassett has an interesting pillared and timbered market hall at its centre.

GARAGES
Artdeans
207 Rodbourne Rd, Swindon
(Hon Yam Ves Lam)
☎(0793) 34985
Rodbourne Service Station
Rodbourne Rd, Swindon
☎(0793) 33168
County
Unit 4/6, Hawksworth Est,
Swindon ☎(0793) 31464
Days Garage & Breakdown Recovery
Cambria Bridge Rd, Swindon
☎(0793) 21465

Junction 17

Chippenham, although largely redeveloped, retains some of its old buildings including Yelde Hall – late 15th-/early 16th-century – which now houses a museum. To the west Corsham has a street

of 16th-century weavers' cottages and Corsham Court, of Elizabethan origin, is open to the public. Sheldon Manor, also to the west, is medieval and has been a family home for some 700 years.

WHERE TO STAY
★★★**Bell House**
Sutton Benger
☎(0249) 720401
FH Mrs C M Parfitt, **Angrove Farm**
Rodbourne ☎(06662) 2982

WHERE TO EAT
PS **Kingston Langley**
Transit Picnic Sites, East and West
2m N of Chippenham
OS173 ST9176

GARAGES
Stanton St Quinton
Stanton St Quinton
☎(06663) 223
D & C Fry
Kington Langley
☎(024975) 228
Corston
(Budd & Chandler) Corston
(Frd) ☎(06662) 3317

The lovely Vale of Kennet forms a major part of the Berkshire countryside. Here the river is shown near the town of Hungerford, on its way to join the Thames near Reading.

Hungerford is well-known for its many antique shops and customers travel from far and wide to browse around its shops, and markets. It is an attractive town, too, with a wide main street.

Newbury is particularly famous for its racecourse where meetings regularly take place. Many trainers have their headquarters nearby, particularly at Lambourne and Kingsclere.

LOCAL RADIO STATIONS	Medium Wave		VHF/FM
	Metres	kHz	MHz
BBC Radio Oxford	202	1458	95.2
IBA Wiltshire Radio			
Swindon area	258	1161	96.4
West Wiltshire	320	936	97.4

M4 Theale - Chippenham

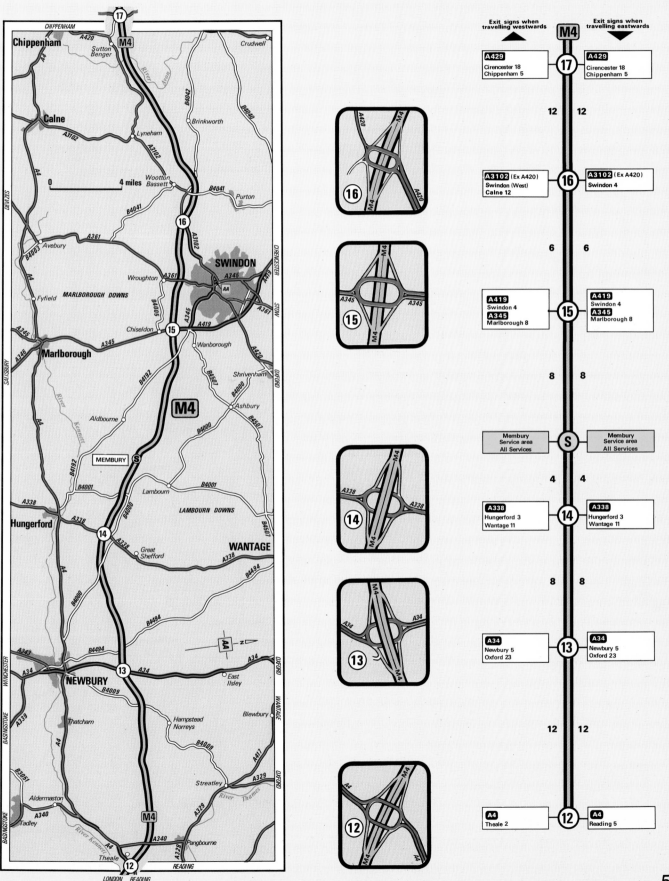

Exit signs when travelling westwards ◄ | M4 | ► Exit signs when travelling eastwards

A429 Cirencester 18 Chippenham 5 — 17 — **A429** Cirencester 18 Chippenham 5

12 — 12

A3102 (Ex A420) Swindon (West) Calne 12 — 16 — **A3102** (Ex A420) Swindon 4

6 — 6

A419 Swindon 4 **A345** Marlborough 8 — 15 — **A419** Swindon 4 **A345** Marlborough 8

8 — 8

Membury Service area All Services — S — Membury Service area All Services

4 — 4

A338 Hungerford 3 Wantage 11 — 14 — **A338** Hungerford 3 Wantage 11

8 — 8

A34 Newbury 5 Oxford 23 — 13 — **A34** Newbury 5 Oxford 23

12 — 12

A4 Theale 2 — 12 — **A4** Reading 5

53

Chippenham - Cardiff M4

Bristol Parkway M32

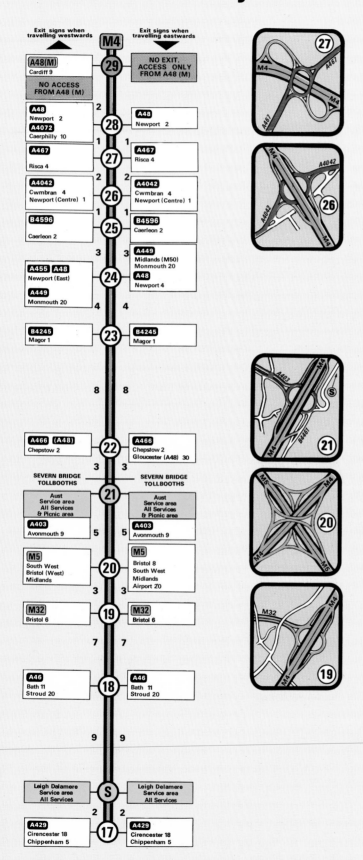

Exit signs when travelling westwards

Exit signs when travelling eastwards

M4

29
A48(M) Cardiff 9
NO ACCESS FROM A48 (M)

NO EXIT. ACCESS ONLY FROM A48 (M)

28
A48 Newport 2
A4072 Caerphilly 10

A48 Newport 2

2

1 | 1

27
A467 Risca 4

A467 Risca 4

2 | 2

26
A4042 Cwmbran 4 Newport (Centre) 1

A4042 Cwmbran 4 Newport (Centre) 1

1 | 1

25
B4596 Caerleon 2

B4596 Caerleon 2

3 | 3

24
A455 A48 Newport (East)
A449 Monmouth 20

A449 Midlands (M50) Monmouth 20
A48 Newport 4

4 | 4

23
B4245 Magor 1

B4245 Magor 1

8 | 8

22
A466 (A48) Chepstow 2

A466 Chepstow 2 Gloucester (A48) 30

3 | 3

SEVERN BRIDGE TOLLBOOTHS

SEVERN BRIDGE TOLLBOOTHS

21
Aust Service area All Services & Picnic area
A403 Avonmouth 9

Aust Service area All Services & Picnic area
A403 Avonmouth 9

5 | 5

20
M5 South West Bristol (West) Midlands

M5 Bristol 8 South West Midlands Airport 20

3 | 3

19
M32 Bristol 6

M32 Bristol 6

7 | 7

18
A46 Bath 11 Stroud 20

A46 Bath 11 Stroud 20

9 | 9

S
Leigh Delamere Service area All Services

Leigh Delamere Service area All Services

2 | 2

17
A429 Cirencester 18 Chippenham 5

A429 Cirencester 18 Chippenham 5

27

26

21

20

19

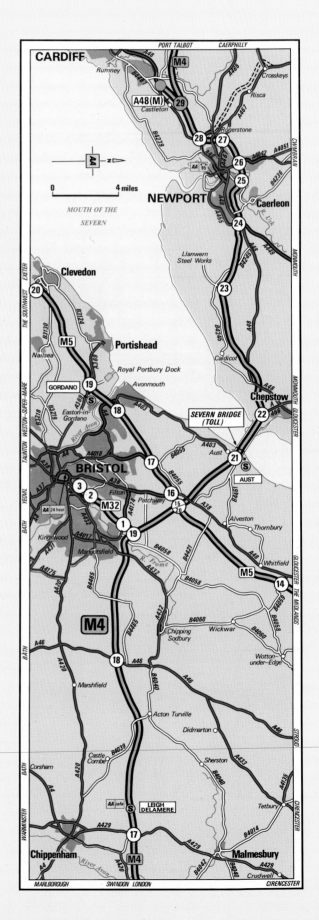

Leigh Delamere Services
(Granada) ☎(0663) 691
Restaurant. Shop.
Playground. Petrol.
Diesel. Breakdowns.
Repairs. HGV parking.
Long-term/overnight
parking for caravans
£3.50. Baby-changing.
For disabled: toilets.
Footbridge. Credit cards
– shop, restaurant,
garage.

Junction 18

Of the nearby large
estates, the grandest must
be Badminton. Dyrham
Park lies to the south.

WHERE TO STAY
★★**Compass Inn**
Tormarton (IH/MINO)
☎(045421) 242
★★**Cross Hands**
Old Sodbury
☎(0454) 313000
GH **Moda**
1 High St, Chipping Sodbury
☎(0454) 312135

WHERE TO EAT
✕**Cordon Bleu**
66 Rounceval St, Chipping
Sodbury ☎(0454) 318041
Vittles Bar
Compass Inn, Tormarton
☎(045421) 242
PS **Tog Hill**
½m E of junction A420/A46.
(6'6" height limit).
OS172 ST7372

GARAGES
Roman Camp
Stroud Rd, Chipping Sodbury
☎(0454) 314202
TT Motors
Hatters Ln, Chipping Sodbury
(BL) ☎(0454) 313181

Junction 19

The M32 branches off here
to Bristol – an interesting
city with a great maritime
tradition. Visit SS Great
Britain, the first iron ship,
the Bristol Industrial
Museum and the National
Lifeboat Museum.

M32 Bristol Parkway

Junction 1

WHERE TO STAY
☆☆☆**Crest**
Filton Rd, Hambrook
☎(0272) 564242

GARAGES
Frenchay Park Service Station
213 Frenchay Park Rd,
Frenchay ☎(0272) 568701
Gold Star Motorcycle Service
22 The Green, Stoke Gifford
☎(0272) 506681
Overndale Service Station
117 Downend Rd, Fishponds
☎(0272) 566846
Streamside
The Stream, Hambrook
☎(0272) 571197

TT Motorcycles
2a Downend Rd, Fishponds
☎(0272) 659690

Junction 2

WHERE TO STAY
GH **Alcove**
508–510 Fishponds Rd,
Fishponds ☎(0272) 653886
GH **Cavendish House**
18 Cavendish Rd, Henleaze
☎(0272) 621017

GARAGES
Auto Safety Centre
140 Ashley Down Rd
☎(0272) 514622
Clarke Bros
175–185 Muller Rd
☎(0272) 513333
J Dangerfield
Staple Hill Rd, Fishponds
☎(0272) 656373
Speedwell
(Hurleyhouse) 7–9
Speedwell Rd
☎(0272) 516971
Terminus Service Station
782–6 Fishponds Rd,
Fishponds ☎(0272) 653164
Williams Automobiles
16–20 Fishponds Rd,
Eastville ☎(0272) 510252

Junction 3

WHERE TO STAY
★★★★**B Grand**
Broad St, Bristol
☎(0272) 291645
☆☆☆**Holiday Inn**
Lower Castle St
☎(0272) 294281
☆☆☆**L Ladbroke Dragonara**
Redcliffe Way
☎(0272) 20044
☆☆☆**Unicorn**
Prince St ☎(0272) 294811
★★★**Avon Gorge**
Sion Hill, Clifton
☎(0272) 738955
★★**Hawthorn's**
Woodland Rd, Clifton
☎(0272) 738432
★★**St Vincent's Rocks**
Sion Hill, Clifton
☎(0272) 739251
GH **Alandale Hotel**
Tyndall's Park Rd, Clifton
☎(0272) 735407
GH **Alexander Hotel**
250–252 Wells Rd, Knowle
☎(0272) 778423
GH **Birkdale Hotel**
11 Ashgrove Rd, Redland
☎(0272) 733635
GH **Cambridge Hotel**
Redland Rd ☎(0272) 736020
GH **Hotel Clifton**
St Pauls Rd, Clifton
☎(0272) 736882
GH **Glenroy Hotel**
30 Victoria Sq, Clifton
☎(0272) 739058
GH **Oakdene**
45 Oakfield Rd
☎(0272) 735900
GH **Oakfield Hotel**
52–54 Oakfield Rd, Clifton
☎(0272) 735556

WHERE TO EAT
✕✕✕✕**Harveys**
12 Denmark St
☎(0272) 277665
✕✕**Marco's**
59 Baldwin St ☎(0272) 24869
✕✕**Rajdoot**
83 Park St ☎(0272) 28033
✕✕**Restaurant du Gourmet**
43 Whiteladies Rd
☎(0272) 736230
✕✕**Tearles**
2 Upper Byron Pl, The
Triangle, Clifton
☎(0272) 28314

✕**Compton's**
52 Upper Belgrave Rd, Clifton
☎(0272) 733515
✕**Ganges**
368 Gloucester Rd, Horfield
☎(0272) 45234
✕**Howard's**
1a Avon Cres, Hotwells
☎(0272) 22921
✕**Raj Tandoori**
35 King St ☎(0272) 291132
✕**Rossi's Ristoranti**
35 Princess Victoria St
☎(0272) 730049

GARAGES
Arban Services
185–189 Easton Rd, Easton
☎(0272) 558030
Auto Safety Centre
Midland Rd, Kingsland
Trading Estate
☎(0272) 552977
Cathedral
College Green
☎(0272) 20031
City Bike Service Centre
4A/5 Charles St, St James
☎(0272) 46318

Junction 20

This is the junction which
links with the M5
Birmingham–Exeter
motorway – see pages
58–68 for details. The
places listed below can be
reached by taking the M5
one mile southwards to
leave at junction 16.

WHERE TO STAY
☆☆☆**Post House**
Thornbury Rd, Alveston (THF)
☎(0454) 412521
★★**Alveston House**
Alveston ☎(0454) 415050
GH **Almondsbury Interchange**
6 Gloucester Rd,
Almondsbury
☎(0454) 613206

WHERE TO EAT
✕**Manor House and Junk Shop**
Gaunts Earthcott
☎(0454) 772225
Ship Inn
Post House, Thornbury Rd,
Alveston ☎(0454) 412521

GARAGES
Callicroft Motor Engineering
Callicroft Rd, Patchway
☎(0272) 695409
Patchway Car Centre
Gloucester Rd, Patchway
(Vau Opl) ☎(0272) 694331
Berkley Vale Motor Co
Alveston (BL)
☎(0454) 412207

Junction 21

This junction is
immediately before the
massive bridge which
spans the Severn Estuary –
of which the adjacent
service area has superb
views.

GARAGES
Forge Service Station
(Forge Cars) Elberton (Sko)
☎(0454) 419051
Tockington Service and Repair
Camp Ln, Elberton
☎(0454) 414670
Olveston Central
New Rd, Olveston
☎(0454) 612266
Pilning
Cross Hands Rd, Pilning
☎(04545) 2909

M4 Chippenham–Cardiff   51 miles

Bypassing Bristol on its north side, the motorway crosses the
lovely Severn Bridge (toll) to enter the Principality of Wales.

Aust Services
(Rank) ☎(04545) 2855
Restaurant. Fast food.
HGV cafe. Picnic site.
Shop. Vending
machines. Petrol. Diesel.
Breakdowns. Repairs.
HGV parking. Long-
term/overnight parking
for caravans £3 (max
24hrs). Baby-changing.
For disabled: toilets, lift
NB steps in building.
Credit cards accepted.

Junction 22

Chepstow, with its castle
and racecourse, are to the
north, beyond which
stretches the beautiful
Wye Valley. Beside a loop
in the river, stands the
majestic Tintern Abbey,
with the Forest of Dean
stretching away to the east.

WHERE TO STAY
★★**Beaufort**
Beaufort Square, Chepstow
☎(02912) 5074
★★**H Castle View**
16 Bridge St, Chepstow
☎(02912) 70349
★★**George**
Moor St, Chepstow (THF)
☎(02912) 2365
GH **First Hurdle**
9 Upper Church St, Chepstow
☎(02912) 2189

WHERE TO EAT
PS **Barnets Wood**
2m W of Chepstow
OS162 ST5194
PS **Prysgau Bach**
3m W of Chepstow
OS162 ST4994

GARAGES
Parkwall
Caldicot, Mathern
☎(0291) 422648
Chepstow Service Station
(W Colthart) Chepstow (BL)
☎(02912) 3159
Larkfield
Newport Rd, Chepstow (Frd)
☎(02912) 2861
Bowens
Tutshill, Chepstow (Tal Peu)
☎(02912) 3131

Junction 23

Penhow Castle to the north
is the oldest inhabited
castle in Wales, dating
from the 12th century.

WHERE TO EAT
PS **Wentwood Lodge**
6½m N of Chepstow
OS171 ST4293

GARAGES
Magor Engineers and Contractors
The Mill, Magor
☎(0633) 880335

Junction 24

To the south are
Newport's steel works,
but a more attractive
prospect lies to the north
where the road follows the
lovely meandering River
Usk.

WHERE TO STAY
☆☆**New Inn Motel**
Langstone (AB)
☎(0633) 412426

WHERE TO EAT
PS **Whitehall Picnic Area**
on A449 3m N of junction 24
OS171 ST3894

GARAGES
Newport Motor Co
Ford House, Lee Way
Industrial Est, Newport (Frd)
☎(0633) 278020

Junction 25

Caerleon, to the north, is
an important Roman site
where excavations have
revealed a legionary
fortress dated AD80–100
and an amphitheatre
which could accommodate
6000 people.

WHERE TO STAY
★★**Priory**
High St, Caerleon, Newport
(AB) ☎(0633) 421241

GARAGES
County Motor Services
248 Conway Rd, Newport
☎(0633) 275120
W H Dodson
Johnsey Trading Est, Gaskell
St, Newport ☎(0633) 57264
Precision Tuning Auto Repairs
(Maindee) 55A Archibald St,
Newport ☎(0633) 841421

Junction 26

The road south leads
directly into the centre of
Newport, a town which
grew around its docks,
exporting coal and iron
products from the Gwent
valleys.

WHERE TO STAY
★★**Queens**
19 Bridge St, Newport (AHT)
☎(0633) 62992
★★**Westgate**
Commercial St, Newport
☎(0633) 66244
★★★**Commodore**
Mill Ln, Llanfravon, Cwmbran
☎(06333) 4091
GH **Caerleon House**
Caeran Rd, Newport
☎(0633) 64869

GARAGES
Shaftesbury Park Lex Mead
Shaftesbury St, Newport (BL
DJ LR Rar) ☎(0633) 858451
Motorwell Servicentre
Granville Sq, George St
Bridge, Newport
☎(0633) 64057

Peter Price
Queens Hill, Newport
☎(0633) 51317

Junction 27

To the north east at
Cwmcarn, a 7-mile-long
Scenic Forest Drive has
been set out by the
Forestry Commission to
give spectacular views of
this mountain forest.

WHERE TO EAT
Michaels
39–41 Tredegar St, Risca
☎(0633) 614300

Junction 28

Adjacent to this junction is
Tredegar House, said to be
the finest Restoration
house in Wales. Its
extensive grounds form a
Country Park with many
facilities for visitors.

WHERE TO STAY
★★★★**Celtic Manor**
The Coldra, Newport
☎(0633) 413000
☆☆**Ladbroke**
(& Conferencentre) The
Coldra, Newport (LB)
☎(0633) 412777

GARAGES
J B Volkswagen
Maesglas Industrial Est,
Newport (VW Aud)
☎(0633) 211326

Junction 29

The A48 (M) leads to
Cardiff's eastern suburbs
beside the Rhymney
River.

WHERE TO STAY
☆☆☆**Ladbroke Wentloog Castle**
Castleton (LB)
☎(0633) 680591
☆☆☆**Post House**
Pentwyn Rd, Pentwyn, Cardiff
(THF) ☎(0222) 731212
★★**St Mellons Hotel and Country Club**
St Mellons ☎(0633) 680355
☆☆☆**Inn on the Avenue**
Circle Way East, Llanedeyrn,
Cardiff ☎(0222) 732520

WHERE TO EAT
✕**La Provençal**
779 Newport Rd, Cardiff
☎(0222) 78262

GARAGES
J B (Volkswagen)
Mardy Rd, Rumney, Cardiff
(Aud VW) ☎(0222) 792301

LOCAL RADIO STATIONS

	Medium Wave		VHF/FM
	Metres	*kHz*	*MHz*
BBC Radio Bristol	194	1548	95.5
IBA Radio West	238	1260	96.3
IBA Wiltshire Radio			
Swindon area	258	1161	96.4
West Wiltshire	320	936	97.4
IBA Gwent Broadcasting	230	1305	104.0
IBA CBC	221	1359	96.0

M4 Cardiff – Pont Abraham   45 miles

From the capital of Wales, the motorway continues westwards past industrial Port Talbot and Swansea and on towards the beautiful south-west peninsula.

One of the beauty spots of South Wales – the lovely Gower Peninsula.

Junction 29 – see page 55

Junctions 30–31 (Projected)

Junction 32

To the south is Cardiff, with its historic buildings, fascinating museums and the Rugby Football Mecca of Cardiff Arms Park. A little way north Castell Coch is a restored 13th-century castle.

WHERE TO STAY
★★**Phoenix**
199–201 Fidlas Rd, Llanishan, Cardiff (MINO) ☎(0222) 764615
☆☆**Griffin Inn Motel**
Rudry (3m E on unclassified rd), Caerphilly ☎(0222) 883396

WHERE TO EAT
Yr Ystafell Gymraeg
74 Whitchurch Rd, Cardiff ☎(0222) 42317
Savastano's
302 North Rd, Cardiff ☎(0222) 30270

GARAGES
Rhiwbina Motor
(C Smart) Rhiwbina, Cardiff (BL) ☎(0222) 63232
Llandaff Motor Co
Llantrisant Rd, Cardiff (Ren) ☎(0222) 562345
A1 Auto Repairs
32 Norbury Rd, Fairwater, Cardiff ☎(0222) 568864
Nantgarw
Cardiff Rd, Nantgarw (Ren) ☎(044385) 2154
Crossways Service Station
Wroughton Pl, Ely, Cardiff ☎(0222) 552918

Crossways Service Station
225 Cowbridge Rd West, Cardiff ☎(0222) 552918
Flower
Cowbridge Rd West, Cardiff (Vlo) ☎(0222) 591182
Grosvenor
(RC Motors) Tintern St, Canton, Cardiff ☎(0222) 28766
Powells
Wroughton Pl, Ely Bridge, Cardiff (AR) ☎(0222) 552918

Junction 33 under construction

Junction 34

This junction is right on the Ely Valley and to the south-west the Hensol Forest has lovely walks and picnic places.

GARAGES
Llantrisant Motors
Talbot Green, Llantrisant (BL Suz Hon Puc) ☎(0443) 223324
Tudor
(Ystradowen) Ystradowen ☎(04463) 2422

Junction 35

The industrial estates of eastern Bridgend can be reached via this junction. The River Ewenny skirts the town, on the banks of which stand the ruins of Ewenny Priory.

LOCAL RADIO STATIONS

	Medium Wave		VHF/FM
	Metres	kHz	MHz
IBA CBC	221	1359	96.0
IBA Swansea Sound	257	1170	95.1

GH **Minerva**
52 Esplanade Av, Porthcawl ☎(065671) 2428
GARAGE
Globe
1 Bridgend Rd, Newton, Porthcawl ☎(065671) 2067

Junctions 38 and 39 R

In contrast to the massive steel works of Port Talbot on one side of the motorway is the Margam Park Country Park on the other. Within its 850 acres are the Castle and Abbey ruins, a theatre and gardens.

WHERE TO STAY
★★★**Ladbroke Twelve Knights**
Margam Rd, Margam (2m SE of A48), Port Talbot (LB) ☎(0639) 882381

Junctions 40 and 41 R

Behind the built up area of Port Talbot are steep wooded valleys. Upstream on the River Afan is the Afan Argoed Country Park.

WHERE TO STAY
★★★**Aberafon**
Aberafon Seafront, Port Talbot ☎(0639) 884949

Junctions 42, 43 and 44

There is a break in the motorway here and travellers must use the A48 until it recommences at junction 44. The establishments listed below are off this road.

WHERE TO STAY
★★**Castle**
The Parade, Neath ☎(0639) 3581
★★**Cimla Court Motel**
Cimla Rd, Neath ☎(0639) 3771

GARAGES
Stadium Service Garage
Main Rd, Skewen, Neath (Frd) ☎(0792) 812495
Cimla Service Station
(Cimla Auto Sales) Cimla Rd, Neath ☎(0639) 3272

Junction 36

The busy town of Bridgend is to the south on the Ogmore River. The remains of Newcastle are at the centre of town, while Coity Castle is to the north-east. Sarn Park Service Area is reached via this junction.

WHERE TO STAY
★★*B* **Wyndham**
Dunraven Pl, Bridgend ☎(0656) 2080

GARAGES
T S Grimshaw
Tremains Rd, Bridgend (Fia Lnc) ☎(0656) 2984
Mid Glam Motors Repairs
Derwen Rd, Bridgend (BL) ☎(0656) 3376
Valeford Motor Co
Cowbridge Rd, Bridgend (Frd) ☎(0656) 4281

Sarn Park Services
☎(0656) 67656
Accommodation. 24hr cafeteria. Shop. Playground. Playroom. Petrol. Diesel. Breakdowns. HGV parking. Long-term/overnight parking for caravans £5. Baby-changing. For disabled: toilets. Credit cards accepted.

Junction 37

The holiday resort of Porthcawl is to the south with fine sandy beaches and rocky headlands.

WHERE TO STAY
★**Rose & Crown**
Nottage (2m N B4283), Porthcawl (WT) ☎(065671) 4850
★★*B* **Maid of Sker**
West Rd, Nottage, Porthcawl ☎(065671) 2172
★★**Atlantic**
West Dr, Porthcawl ☎(065671) 5011
★★**Fairways**
Sea Front, Porthcawl ☎(065671) 2085
★*B* **Brentwood**
37/41 Mary St, Porthcawl ☎(065671) 2725
★**Seaways**
26–28 Mary St, Porthcawl ☎(065671) 3510
GH **Collingwood**
40 Mary St, Porthcawl ☎(065671) 2899

WHERE TO STAY
☆☆☆**Heronston**
Ewenny (2m S B4265), Bridgend (IH) ☎(0656) 68811

WHERE TO EAT
×××**Coed y Mwstwr**
Coychurch (3m E Bridgend, 1m NE Coychurch), Bridgend ☎(0656) 860621
×**Riverside**
20 Castle St, Swansea ☎(0792) 55210
××**Oyster Perches**
Uplands Cres, Swansea ☎(0792) 473173

Junction 45

Swansea is the second largest city in Wales and its biggest industrial complex. The Maritime and Industrial Museum includes a working woollen mill and steam locomotives.

GARAGES
Cambrian
(G J Birt & Son) Down St, Clydach ☎(0792) 843222
Siloh Motors
Cwm Level Rd, Landore, Swansea (Suz Sko) (Westbound only) ☎(0792) 43732
McFarlane Motors
Cwmfelin Industrial Est, Heol-y-Gors, Cwmrwria, Swansea ☎(0792) 43289
Pep Auto Services
56–60 Morfa Rd, Strand, Swansea ☎(0792) 468134

Junction 46 R

Another access road to Swansea leads off from this junction.

GARAGES
Siloh Motors
Cwm Level Rd, Landore, Swansea (Suz Sko) (Eastbound only) ☎(0792) 43732

Junction 47

Yet another road into Swansea, but this one also leads to the beautiful Gower, a popular holiday area. The Penllergaer Forest lies to the north of the junction.

WHERE TO STAY
☆☆☆**Fforest Motel**
Pontardulais Rd, Fforestfach (on A438, ½m S of M4 junc 47), Swansea ☎(0792) 588711
FH Mr and Mrs G Davies, **Croft Farm** Heol-y-Barna, Pontardulais ☎(0792) 883654
FH Mr F Jones, **Coynant Farm** Felindre ☎(0269) 5640 & 2064

GARAGES
C E M Day
Llanelli Rd, Gamgoch, Swansea (Frd) ☎(0792) 893041

Central
Sterry Rd, Gowerton ☎(0792) 873249

Junction 48

The River Loughor flows into a wide estuary south of this junction upon which stands the steel-producing town of Llanelli. Parc Howard is the home of the town's Art Gallery and Museum.

WHERE TO STAY
★★**Diplomat**
Ael-y-Bryn, Llanelli ☎(05542) 56156

GARAGES
Yspitty Service Station
(Manton Motors) Heoe y Bwlch, Bynea ☎(05542) 54450

Junction 49

The motorway ends here at Pont Abraham on the River Gwili. Continue on the A48 for Carmarthen and the Pembrokeshire coast.

WHERE TO EAT
PS **Pont Abraham Transit Picnic Site**
Adjacent to motorway. OS159 SN5707

GARAGES
Derlwyn
(Gwyn Williams & Son) Tumble ☎(0269) 841312

Pont Abraham Services
(Roadchef) ☎(0782) 884663
Restaurant 07.00–23.00. Fast food. Picnic area. Tourist information. AA Motoring Information Centre. Vending machines. Petrol. Diesel. HGV parking. Long-term/overnight parking for caravans. Baby-changing. For disabled: toilets, ramp. Credit cards accepted.

Cardiff Castle was restored by the 3rd Marquess of Bute with the help of architect William Burges.

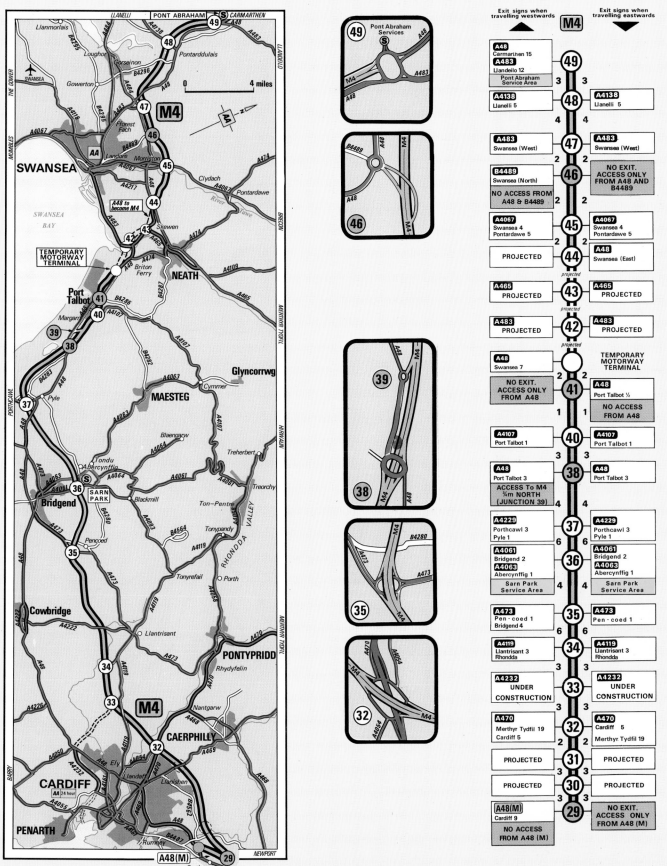

Exit signs when travelling westwards ▲

Exit signs when travelling eastwards ▼

M4

Westwards	Jct	Eastwards
A48 Carmarthen 15 / A483 Llandeilo 12 / Pont Abraham Service Area	49	
	3 / 3	
A4138 Llanelli 5	48	A4138 Llanelli 5
	4 / 4	
A483 Swansea (West)	47	A483 Swansea (West)
	2 / 2	
B4489 Swansea (North) / NO ACCESS FROM A48 & B4489	46	NO EXIT. ACCESS ONLY FROM A48 AND B4489
	2 / 2	
A4067 Swansea 4 / Pontardawe 5	45	A4067 Swansea 4 / Pontardawe 5
	2 / 2	
PROJECTED	44	A48 Swansea (East)
	projected	
A465 PROJECTED	43	A465 PROJECTED
	projected	
A483 PROJECTED	42	A483 PROJECTED
	projected	
A48 Swansea 7 / NO EXIT. ACCESS ONLY FROM A48	○ TEMPORARY MOTORWAY TERMINAL	TEMPORARY MOTORWAY TERMINAL
	2 / 2	
	41	A48 Port Talbot ½ / NO ACCESS FROM A48
	1 / 1	
A4107 Port Talbot 1	40	A4107 Port Talbot 1
	3 / 3	
A48 Port Talbot 3 / ACCESS To M4 ¾m NORTH (JUNCTION 39)	38	A48 Port Talbot 3
A4229 Porthcawl 3 / Pyle 1	37	A4229 Porthcawl 3 / Pyle 1
A4061 Bridgend 2 / A4063 Abercynffig 1 / Sarn Park Service Area	36	A4061 Bridgend 2 / A4063 Abercynffig 1 / Sarn Park Service Area
	4 / 4	
A473 Pen - coed 1 / Bridgend 4	35	A473 Pen - coed 1
	6 / 6	
A4119 Llantrisant 3 / Rhondda	34	A4119 Llantrisant 3 / Rhondda
A4232 UNDER CONSTRUCTION	33	A4232 UNDER CONSTRUCTION
	3 / 3	
A470 Merthyr Tydfil 19 / Cardiff 5	32	A470 Cardiff 5 / Merthyr Tydfil 19
PROJECTED	31	PROJECTED
PROJECTED	30	PROJECTED
A48(M) Cardiff 9 / NO ACCESS FROM A48(M)	29	NO EXIT. ACCESS ONLY FROM A48 (M)

Inset maps:

49 Pont Abraham Services — S

46

39 / 38

35

32

Birmingham - Cheltenham M5

Ross Spur M50

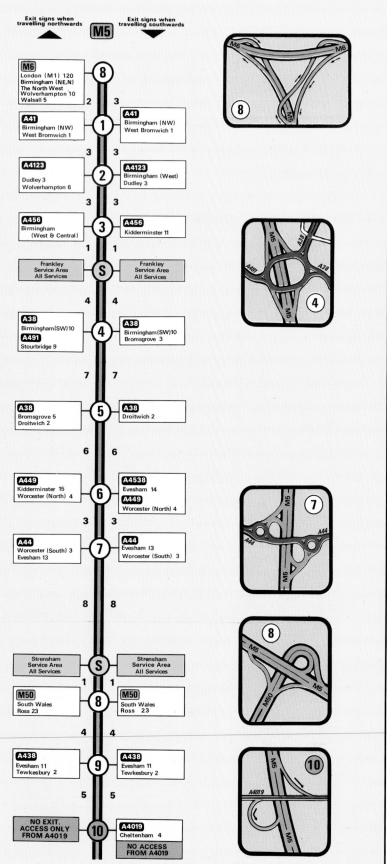

8
M6
London (M1) 120
Birmingham (NE,N)
The North West
Wolverhampton 10
Walsall 5

2   3

1
A41
Birmingham (NW)
West Bromwich 1

A41
Birmingham (NW)
West Bromwich 1

3   3

2
A4123
Dudley 3
Wolverhampton 6

A4123
Birmingham (West)
Dudley 3

3   3

3
A456
Birmingham
(West & Central)

A456
Kidderminster 11

1   1

S
Frankley
Service Area
All Services

Frankley
Service Area
All Services

4   4

4
A38
Birmingham(SW)10
A491
Stourbridge 9

A38
Birmingham(SW)10
Bromsgrove 3

7   7

5
A38
Bromsgrove 5
Droitwich 2

A38
Droitwich 2

6   6

6
A449
Kidderminster 15
Worcester (North) 4

A4538
Evesham 14
A449
Worcester (North) 4

3   3

7
A44
Worcester (South) 3
Evesham 13

A44
Evesham 13
Worcester (South) 3

8   8

S
Strensham
Service Area
All Services

Strensham
Service Area
All Services

1   1

8
M50
South Wales
Ross 23

M50
South Wales
Ross 23

4   4

9
A438
Evesham 11
Tewkesbury 2

A438
Evesham 11
Tewkesbury 2

5   5

10
NO EXIT.
ACCESS ONLY
FROM A4019

A4019
Cheltenham 4

NO ACCESS
FROM A4019

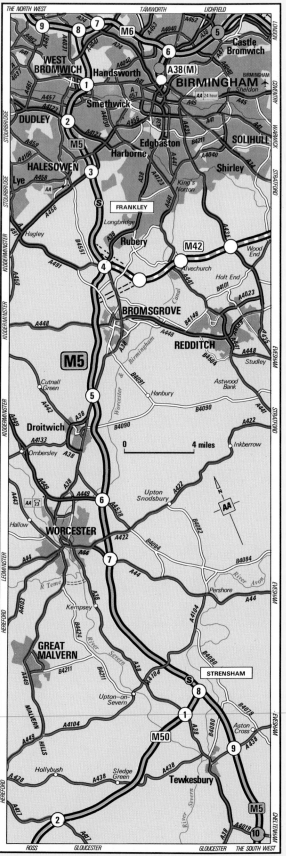

Junction 1

After branching off from the M6, this first junction is at West Bromwich, west of Birmingham's centre.

WHERE TO STAY
☆☆☆**West Bromwich Moat House**
Birmingham Rd, West Bromwich (QM)
☎021–553 6111

GARAGES
Colmore Car People
Birmingham Road, West Bromwich (Fia Fer)
☎021–553 7509
Charles Clark
High St, West Bromwich (BL DJ) ☎021–553 6201

Junction 2

To the north west are Dudley and the Black Country Museum.

WHERE TO STAY
★★**Station**
Birmingham Rd, Dudley (WDB) ☎(0384) 53418
★**Ward Arms**
Birmingham Rd, Dudley (WDB) ☎(0384) 52723
GH **Highfield**
Waterfall Ln, Rowley Regis
☎021–559 1066

WHERE TO EAT
❀××**Jonathans**
16 Wolverhampton Rd, Quinton, Oldbury
☎021–429 3757
×**Franzi's**
151 Milcote Rd, Bearwood, Birmingham ☎021–429 7920

GARAGES
Hanger Motor
4 Wolverhampton Rd, Oldbury (Frd)
☎021–429 7111
Jackson & Lawley
9 Brades Rd, Oldbury
☎021–552 1737
Kissane & Jeffries
Station Rd, Rowley Regis
☎021–559 1085

Junction 3

The Birmingham suburb of Edgbaston is to the east.

WHERE TO STAY
★★**Hotel Annabelle**
19 Sandon Rd, Edgbaston
☎021–429 1182
☆☆**Apollo**
243–247 Hagley Rd, Edgbaston ☎021–455 0271
❀★★★★B **Plough & Harrow**
Hagley Rd, Edgbaston (CRH)
☎021–454 4111
★★**Cobden**
166 Hagley Rd, Edgbaston, Birmingham ☎021–454 6621
GH **Wentworth**
103 Wentworth Rd, Harborne
☎021–427 2839

WHERE TO EAT
×**Michelle**
182–184 High St, Harborne
☎021–426 4133

GARAGES
Five Star Motors
Hagley Rd, Halesowen (Hon Sab) ☎021–550 6416
Patrick Motors
Hagley Rd West, Halesowen (BL) ☎021–422 7171

Frankley
(Granada)
☎021–550 3131
Restaurant. Fast food. Shop. Petrol. Diesel. Breakdowns. Repairs. HGV parking. Long-term/overnight parking for caravans (£3.50 per night). Baby-changing. For disabled: toilets, **NB** steps within building. Credit cards – shop, garage, restaurant.

Junction 4

Several country parks are near this junction.

WHERE TO STAY
★★★**Perry Hall**
Kidderminster Rd, Bromsgrove (EH)
☎(0527) 31976

WHERE TO EAT
××**Bell Inn**
Bell End, Belbroughton
☎(0562) 730232
❀×××**Grafton Manor**
Grafton Rd, Bromsgrove
☎(0527) 31525
PS **Cofton Common**
3m NE of Bromsgrove
OS139 SP0076

GARAGES
Clarkes Motor Services
Lickey Rd, Rednall (BL)
☎021–453 7127
Tessall
1306 Bristol Rd South, Longbridge (Ren)
☎021–475 5241
Holy Cross
Bromsgrove Rd, Holy Cross, Clent (Fia) ☎(0562) 730557

Junction 5

A 15-acre site at Stoke Heath, to the north east, contains the Avoncroft Museum of Buildings.

WHERE TO STAY
★★★★H **Château Impney**
Droitwich ☎(0905) 774411
★★★★**Raven**
St Andrew's St, Droitwich
☎(0905) 772224
★**St Andrew's House**
Worcester Rd, Droitwich
☎(0905) 773202

WHERE TO EAT
×**Spinning Wheel Restaurant**
13 St Andrew's St, Droitwich
☎(0905) 770031

GARAGES
Wychbold
Worcester Rd, Wychbold
☎(052 786) 448

Junction 6

The city of Worcester is famous for its china and its fine cathedral.

WHERE TO STAY
★**Park House**
12 Droitwich Rd, Worcester
☎(0905) 21816
★**Talbot**
8–10 Barbourne Rd, Worcester ☎(0905) 21206
FH Mrs P Wardle, **Pear Tree's Farm**, Oddingley
☎(0905) 778489

GARAGES
Warndon Service Station
Cranham Dr, Warndon
☎(0905) 52970
Colmore Car People
Old Brewery Service Station, 21 Barbourne Rd, Worcester
☎(0905) 28461
Top Gear
391 Ombersley Rd, Worcester (Dhu)
☎(0905) 53034

Junction 7

This is the southern access to Worcester.

WHERE TO STAY
★★★**Giffard**
High St, Worcester (THF)
☎(0905) 27155
★★**Star**
Foregate St, Worcester (WDB) ☎(0905) 24308
GH **Loch Ryan**
119 Sidbury Rd, Worcester
☎(0905) 351143
INN **Five Ways**
14 Angel Place, Worcester
☎(0905) 27065

WHERE TO EAT
❀××**Brown's**
The Old Cornmill, South Quay, Worcester
☎(0905) 26263
××**King Charles II**
New St, Worcester
☎(0905) 22449
×**Purbani Tandoori**
27 The Tything, Worcester
☎(0905) 27402

GARAGES
Colourcraft
Unit 9a, Shrubhill Industrial Est, Worcester
☎(0905) 24493
Law H A J
New Trading Est, Newton Rd, Worcester ☎(0905) 21366
Tolladine Service Station
Tolladine Rd, Worcester
☎(0905) 24213
Severn Bridge Service Station
(General Motor Services)
New Rd, Worcester
☎(0905) 421028

The central tower of Worcester Cathedral was completed in 1374 and commands a fine view of the lovely Malvern Hills. It has an attractive riverside setting.

M5 Birmingham–Cheltenham   48 miles

From England's second largest city, the motorway follows the lovely Severn Valley southwards.

Strensham Services
(Kennings)
☎(0684) 293004
24hr Cafeteria. Waitress Service (north side) 09.00–21.00. Shop. Vending machines. Petrol. Diesel. Breakdowns. Repairs. HGV parking. Overnight parking for caravans £3. For disabled: toilets, ramp, **NB** steps in building. Footbridge. Roadbridge. Credit cards – garage.

Junction 8

The M50 branches off here for Ross-on-Wye. See below for details.

Junction 9

Tewkesbury is a delightful town with many old half-timbered buildings.

WHERE TO STAY
★★★**Bell**
Church St, Tewkesbury
☎(0684) 293293
☆☆B **Tewkesbury Park Golf and Country Club**
Lincoln Green Ln, Tewkesbury ☎(0684) 295405
★B **Tudor House**
High St, Tewkesbury
☎(0684) 297755
GH **Ancient Grudge**
15 High St, Tewkesbury
☎(0684) 292204

GARAGES
Graham Wright Motors
Ashchurch Rd, Tewkesbury (Frd) ☎(0684) 292398
Mitton Manor
(Warners) Bredon Rd, Tewkesbury (BL Peu Tal)
☎(0684) 293122
Warners
Gloucester Rd, Tewkesbury (BL Peu Tal) ☎(0684) 293122
Wicliffe Motor Co
90 Gloucester Rd, Tewkesbury (BL)
☎(0684) 292370

Junction 10 R

Southbound travellers* can leave the motorway here for Cheltenham. *Northbound travellers see junction 11 on page 60.

WHERE TO STAY
★**Royal Ascot**
Western Rd, Cheltenham
☎(0242) 513640

★★★★**Queen's**
The Promenade, Cheltenham (THF) ☎(0242) 514724
★★★**Carlton**
Parabola Rd, Cheltenham
☎(0242) 514453
★★**George**
St George's Rd, Cheltenham
☎(0242) 35751

WHERE TO EAT
×**Rajvooj Tandoori**
1 Albion St, Cheltenham
☎(0242) 24288
×**Mister Tsang**
63 Winchcombe St, Cheltenham ☎(0242) 38727

GARAGES
Lex Mead
Princess Elizabeth Way, Cheltenham (MG BL DJ Rar)
☎(0242) 20441
Mann Egerton
Imperial House, Montpellier Spa Rd, Cheltenham (Ren)
☎(0242) 21651
Bristol Street Motors
83–93 Winchcombe St, Cheltenham (Frd)
☎(0242) 27061

LOCAL RADIO STATIONS

	Medium Wave		VHF/FM
	Metres	kHz	MHz
BBC Radio WM	206	1458	95.6
IBA BRMB Radio	261	1152	94.8
IBA Beacon Radio	303	990	97.2
IBA Radio Wyvern			
Hereford area	314	954	95.8
Worcester area	196	1530	96.2
IBA Severn Sound	388	774	95.0

M50 Ross Spur   22 miles

This motorway leads to Ross-on-Wye, beyond which South Wales is easily accessible.

Junction 1

This junction, close to the River Severn, leads to Tewkesbury – see M5 junction 9 for details.

WHERE TO STAY
★★**White Lion**
High Street, Upton upon Severn ☎(06846) 2551

WHERE TO EAT
××**Gay Dog Inn**
Baughton ☎(06846) 2706
Cromwells
16–18 Church St, Upton upon Severn ☎(06846) 2447

GARAGES
Stratford Bridge
Ripple ☎(06846) 2657
Grove House Motors
Ryall ☎(06846) 3234
Ryall
Tewkesbury Rd, Upton upon Severn ☎(06846) 2271

Junction 2

Ledbury to the north is a lovely little town with lots of old timbered buildings.

WHERE TO STAY
★★**Feathers**
High St, Ledbury
☎(0531) 2600

★**Royal Oak**
The Southend, Ledbury
☎(0531) 2110

WHERE TO EAT
Applejack
44 the Homend, Ledbury
☎(0531) 4181

GARAGES
G Hopkins & Sons
New St, Ledbury
☎(0531) 2233
Parkway
(Gittings Bros Motors)
Ledbury ☎(0531) 2320

Junction 3

To the east, near Newent, is the Falconry Centre.

WHERE TO STAY
★★**How Caple Grange**
How Caple ☎(098986) 208

WHERE TO EAT
×**Soutters**
Culver St, Newent
☎(0531) 820896

GARAGES
Westons
Much Marcle ☎(053184) 232
Lea
Lea ☎(098981) 295

Junction 4

Standing above a horseshoe bend in the lovely River Wye is the town of Ross-on-Wye.

WHERE TO STAY
★★★H **Chase**
Gloucester Rd, Ross-on-Wye
☎(0989) 63161
★★**King's Head**
8 High Street, Ross-on-Wye
☎(0989) 63174
★**Brookfield House**
Over Ross, Ross-on-Wye
☎(0989) 62188
★**Rosswyn**
17 High St, Ross-on-Wye
☎(0989) 62733
GH **Bridge House**
Wilton, Ross-on-Wye
☎(0989) 62655
GH **Ryefield House**
Gloucester Rd, Ross-on-Wye
☎(0989) 63030

WHERE TO EAT
The Old Pheasant
52 Edde Cross St, Ross-on-Wye ☎(0989) 65751

GARAGES
T C Longford
Cantilupe Rd, Ross-on-Wye
☎(0989) 62400
Overross
Overross St, Ross-on-Wye
☎(0989) 63666

M5 Cheltenham – Clevedon   49 miles

With the Cotswolds to the east and the widening River Severn to the west, the motorway continues southwards.

Junction 10 see page 59

Junction 11

The local airport serving Gloucester and Cheltenham is close to this junction. Gloucester has many fine old buildings and interesting museums and its Cathedral is one of the glories of British architecture.

WHERE TO STAY
☆☆☆☆ **Golden Valley Thistle**
Gloucester Rd, Cheltenham (TS) ☎(0242) 32691
✹★★★ **Greenway**
Shurdington, Cheltenham ☎(0242) 862352
☆☆☆ **Crest**
Crest Way, Barnwood, Gloucester (CRH) ☎(0452) 63311
★★★ **Carlton**
Parabola Rd, Cheltenham ☎(0242) 514453
★★★ **B Wyastone**
Parabola Rd, Cheltenham (EXEC) ☎(0242) 22659
★★ **George**
St George's Rd, Cheltenham ☎(0242) 35751
★ **Royal Ascot**
Western Rd, Cheltenham ☎(0242) 513640
★★★★ **Queen's**
The Promenade, Cheltenham (THF) ☎(0242) 514724
GH **Beaumont House**
56 Shurdington Rd, Cheltenham ☎(0242) 45986
GH **Willoughby**
1 Suffolk Sq, Cheltenham ☎(0242) 22798
GH **Wellington**
Wellington Sq, Cheltenham ☎(0242) 21627
GH **Cleevelands House**
38 Evesham Rd, Cheltenham ☎(0242) 518898

WHERE TO EAT
✕ **Mister Tsang**
63 Winchcombe St, Cheltenham ☎(0242) 38727
✕ **Rajvooj Tandoori**
1 Albion St, Cheltenham ☎(0242) 24288
Forrest's Wine Bar
Imperial Ln, Cheltenham ☎(0242) 38001
Montpellier Wine Bar and Bistro
Bayshill Lodge, Montpellier St, Cheltenham ☎(0242) 27774

GARAGES
Twynings Service Station
(Wren Car Sales) Main Rd, Shurdington ☎(0242) 862234
Lex Mead
Princess Elizabeth Way, Cheltenham (MG BL DJ Rar) ☎(0242) 20441
Cleevely Motors
17 Andover St, Cheltenham ☎(0242) 21988
Naunton Park
(New Victory Mechanics) Churchill Rd, Cheltenham ☎(0242) 26979
Twigworth Services
(Robert Bamford) Tewkesbury Rd, Twigworth ☎(0452) 730070
Bristol Street Motors
83–93 Winchcombe St, Cheltenham (Frd) ☎(0242) 27061

Mann Egerton
Imperial House, Montpellier Spa Rd, Cheltenham (Ren) ☎(0242) 21651
Pihlens Motors
60–66 Fairview Rd, Cheltenham (Dat) ☎(0242) 513020
Yeats
80–86 Prestbury Rd, Cheltenham ☎(0242) 580660

Junction 12 R

Southbound travellers cannot leave the motorway here, but those travelling northwards have access to the Gloucester road. 2 miles south of the city is the Robinswood Hill Country Park.

WHERE TO STAY
★★★ **HBL Painswick**
Kemps Ln, Painswick ☎(0452) 812160

WHERE TO EAT
PS **Robinswood Hill**
2m S of Gloucester. OS162 SO8315

GARAGES
Skipper
Cole Av, Gloucester (Vau Opl) ☎(0452) 26711

Junction 13

Beside the River Severn to the south west is Slimbridge, Sir Peter Scott's famous wildfowl Trust which includes the world's largest collection of swans, geese and ducks.

WHERE TO STAY
★★★ **Stonehouse Court**
Bristol Rd, Stonehouse ☎(045382) 5155

WHERE TO EAT
The Old Forge
Whitminster ☎(0452) 740875
The Ship
Post House Hotel, Thornbury Rd, Alveston ☎(0454) 412521
PS **Coaley Park**
2½m N of Uley OS162 SO7901

GARAGES
Frombridge
Whitminster ☎(0452) 740979
King's Stanley
King's Stanley ☎(045382) 2105

Michael Wood Services
(Welcome Break) ☎(0454) 260631
Restaurant. Picnic area. Shop. Vending machines. Petrol. Diesel. Liquid Petroleum Gas. Breakdowns. HGV parking. Long-term/overnight parking for caravans £3. Baby-changing. For disabled: toilets. Footbridge.

Junction 14

Berkeley Castle, to the north, has been the home of the Berkeley family for over 800 years and was the place where Edward II was murdered in 1327. Nearby

the Jenner Museum commemorates the man who discovered smallpox vaccine.

WHERE TO STAY
★★ **Park**
Falfield ☎(0454) 260550
☆☆ **Newport Towers Motel**
(on A38) Newport, Glos ☎(0453) 810575
✹★★★ **Thornbury Castle**
Thornbury (PRE) ☎(0454) 412647
GH **Elms**
Stone ☎(0454) 260279
FH Mr & Mrs Bryant **Green Farm** Falfield ☎(0454) 260319
FH Mrs S Scolding **Varley Farm** Talbot End, Cromhall ☎(045424) 292

GARAGES
Whitfield
(H C Sams & Son) Falfield ☎(0452) 260296
Taylors of Woodford
Woodford (A38), Berkeley (Frd) ☎(0454) 260133
Maar (International) Supplies
Gloucester Rd, Grovesend (LR Rar) ☎(0454) 413083

Junction 15

This is the junction with the M4 – South Wales to the west and London to the east. See pages 50–57 for details.

Junction 16

The A38 south leads into Bristol on its northern side. Details of additional hotels, restaurants and garages in Bristol can be found on page 55.

WHERE TO STAY
★★★ **Grange at Northwoods**
Northwoods, Winterbourne ☎(0454) 777333
☆☆ **Post House**
Thornbury Rd, Alveston (THF) ☎(0454) 412521
★★ **Alveston House**
Alveston ☎(0454) 415050
☆☆☆ **Crest**
Filton Rd, Hambrook, Bristol (CRH) ☎(0272) 564242
GH **Almondsbury Interchange**
6 Gloucester Rd, Almondsbury ☎(0454) 613206

WHERE TO EAT
✕ **Manor House and Junk Shop**
Gaunts Earthcott ☎(0454) 772225

GARAGES
Callicroft Motor Engineering
Callicroft Rd, Patchway ☎(0272) 695409
Patchway Car Centre
Gloucester Rd, Patchway (Vau Opl) ☎(0272) 694331
Gold Star Motorcycle Service
22 The Green, Stoke Gifford, Bristol ☎(0272) 793794
Olveston Central
New A38, Olveston ☎(0454) 612266
Berkley Vale Motor Co
Alveston (BL) ☎(0454) 412207

Junction 17

The airfield to the south belongs to the aircraft works at Filton. At Henbury is Blaise Castle House, now a social history museum.

WHERE TO STAY
GH **Westbury Park**
37 Westbury Rd, Westbury-on-Trym ☎(0272) 620465

WHERE TO EAT
✕ **Ganges**
368 Gloucester Rd, Horfield ☎(0272) 45234
✕ **Howards**
1a Avon Cres, Hotwells ☎(0272) 22921
✕ **Comptons**
52 Upper Belgrave Rd, Clifton ☎(0272) 733515

GARAGES
Beaufort Service Station
(Auto Motor Recovery) Wyck Beck Rd, Henbury ☎(0272) 500282
Henleaze Park Service Station
144 Henleaze Rd, Westbury-on-Trym ☎(0272) 622608
Wellington Service Station
(A J Waycott) Wellington Hill West, Westbury-on-Trym ☎(0272) 45641
Pilning
Cross Hands Rd, Pilning ☎(04545) 2909

Junction 18

The road into Bristol follows the River Avon to the famous Clifton Suspension Bridge which spans the Avon Gorge. West of the junction are the Avonmouth Docks.

WHERE TO STAY
★★★ **Avon Gorge**
Sion Hill, Clifton (MC) ☎(0272) 738955
★★★ **St Vincent's Rocks**
Sion Hill, Clifton ☎(0272) 739251

GARAGES
Flyover Services
West Town Rd, Avonmouth ☎(0272) 823037
AFS Garages
St Andrews Rd, Avonmouth (Frd) ☎(0272) 822983
Pembroke Service Station
High St, Shirehampton ☎(0272) 827396

Junction 19

Portishead is to the west, and Gordano Services are located at this junction.

WHERE TO STAY
☆☆ **Redwood Lodge**
Beggar Bush Ln, Failand ☎(027580) 3901

WHERE TO EAT
Peppermill Restaurant
3 The Precinct, Portishead. ☎(0272) 847407

GARAGES
Portishead Service Station
(Clist & Rattle) 120 High St, Portishead (BL) ☎(0272) 843444
Station Ford
Cabstand, Portishead (Frd) ☎(0272) 842180
Cambridge Batch
(Clist & Rattle Ltd) Flax Bourton (BL RT) ☎(027583) 3666

Gloucester Cathedral is one of the centres of the famous Three Choirs Festival. It is centred on a great Norman church, but was later transformed into a fine example of the Perpendicular style.

Gordano Services
(THF) ☎(027583) 3074
24hr Cafeteria. Picnic area. Shop. Playground. Petrol. Diesel. Breakdowns. Repairs. Long-term/overnight parking for caravans £2.50. Baby-changing. For disabled: ramp. Credit cards – shop, garage, restaurant.

Junction 20

The 14th-century manor house of Clevedon Court has associations with Tennyson and Thackeray. The nearby Clevedon Craft Centre consists of 14 craft studios of various kinds.

WHERE TO STAY
★★★ **Walton Park**
Wellington Ter, Clevedon ☎(0272) 874253
GH **Amberley**
146 Old Church Rd, Clevedon ☎(0272) 874402
FH K Dee-Shapland **Green Farm** Claverham ☎(0934) 833180

INN **Star**
Tickenham ☎(0272) 852071
INN **Prince of Orange**
High St, Yatton ☎(0934) 832193

WHERE TO EAT
Mon Plaisir Restaurant
32–34 Hill Road, Clevedon ☎(0272) 872307

GARAGES
Kenn
Kenn ☎(0272) 876339
Binding & Payne
Old Church Rd, Clevedon (BL) ☎(0272) 872201
Clevedon
Bristol Rd, Clevedon (Tal Frd) ☎(0272) 873701
Parnell Rd
Parnell Rd, Clevedon ☎(0272) 874576
Wayside
(P Neill) Kenn, Clevedon ☎(0272) 874119
Tickenham
Clevedon Rd, Tickenham ☎(0272) 852035
Gordano Motor Services
Clevedon Rd, Weston-in-Gordano ☎(0272) 842690
Autospeed Tyre & Exhaust Centre
Hannah More Rd, Nailsea ☎(0272) 855247

LOCAL RADIO STATIONS

	Medium Wave		VHF/FM
	Metres	kHz	MHz
BBC Radio Bristol	194	1548	95.5
IBA Radio West	238	1260	96.3
IBA Severn Sound	388	774	95.0

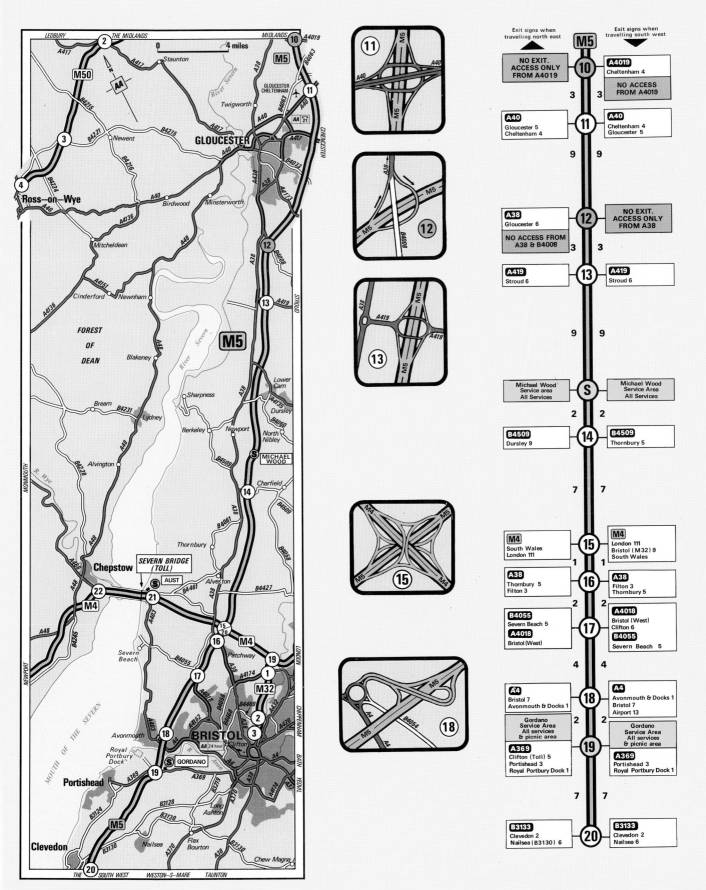

Clevedon - Waterloo Cross M5

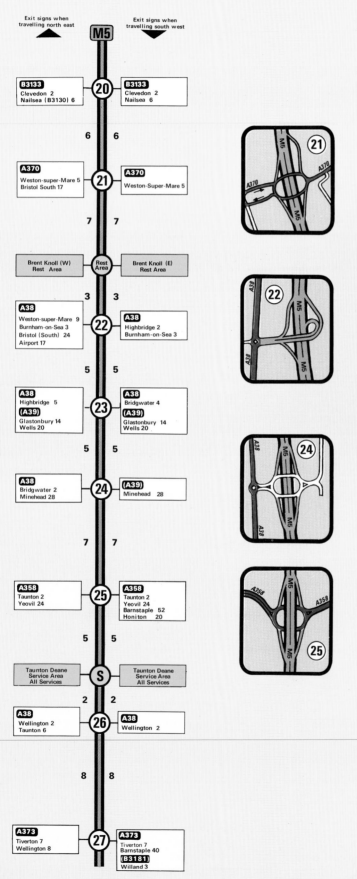

Exit signs when travelling north east

Exit signs when travelling south west

M5

20

B3133
Clevedon 2
Nailsea (B3130) 6

B3133
Clevedon 2
Nailsea 6

6   6

21

A370
Weston-super-Mare 5
Bristol South 17

A370
Weston-Super-Mare 5

7   7

Rest Area

Brent Knoll (W)
Rest Area

Brent Knoll (E)
Rest Area

3   3

22

A38
Weston-super-Mare 9
Burnham-on-Sea 3
Bristol (South) 24
Airport 17

A38
Highbridge 2
Burnham-on-Sea 3

5   5

23

A38
Highbridge 5
(A39)
Glastonbury 14
Wells 20

A38
Bridgwater 4
(A39)
Glastonbury 14
Wells 20

5   5

24

A38
Bridgwater 2
Minehead 28

(A39)
Minehead 28

7   7

25

A358
Taunton 2
Yeovil 24

A358
Taunton 2
Yeovil 24
Barnstaple 52
Honiton 20

5   5

S

Taunton Deane
Service Area
All Services

Taunton Deane
Service Area
All Services

2   2

26

A38
Wellington 2
Taunton 6

A38
Wellington 2

8   8

27

A373
Tiverton 7
Wellington 8

A373
Tiverton 7
Barnstaple 40
(B3181)
Willand 3

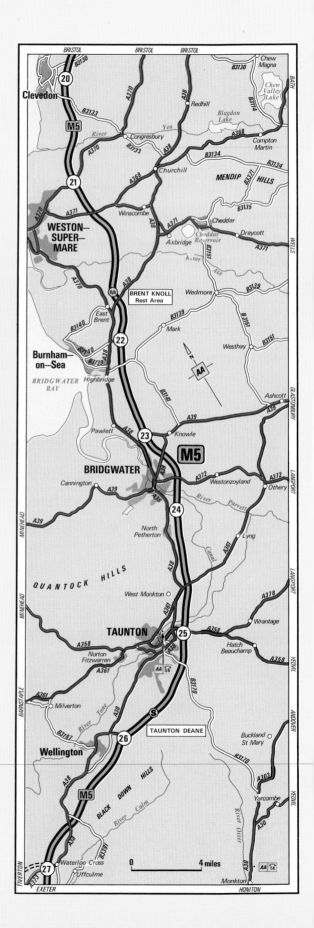

0   4 miles

AA 14

Junction 21

Much holiday traffic is likely to be heading for this junction which lies between Weston-super-Mare and the lovely Mendip Hills. Cheddar is not far to the south east.

WHERE TO STAY
★★**Grand Panorama**
57 South Rd, Weston-super-Mare ☎(0934) 24980
★★★*L* **Grand Atlantic**
Beach Rd, Weston-super-Mare (THF) ☎(0934) 26543
★**Monte Bello**
48 Knightstone Rd, Weston-super-Mare ☎(0934) 22303
★★★*H* **Royal Pier**
Birnbeck Rd, Weston-super-Mare ☎(0934) 26644
GH **Kara**
Hewish ☎(0934) 834442
GH **Southmead**
435 Locking Rd, Weston-super-Mare ☎(0934) 29351
GH **Lydia**
78 Locking Rd, Weston-super-Mare ☎(0934) 25962
GH **Newton House**
79 Locking Rd, Weston-super-Mare ☎(0934) 29331
GH **Shire Elms**
71 Locking Rd, Weston-super-Mare ☎(0934) 28605
GH **Wychwood**
148 Milton Rd, Weston-super-Mare ☎(0934) 27793
GH **Fourways**
2 Ashcombe Rd, Weston-super-Mare ☎(0934) 23827

WHERE TO EAT
××**Sands**
17 Beach Rd, Weston-super-Mare ☎(0934) 414414
Chris's Restaurant
8 Alexandra Parade, Weston-super-Mare ☎(0934) 23481

GARAGES
Passey & Porter
Locking Rd, Weston-super-Mare ☎(0934) 28291
Sandford Service Station
Greenhill Rd, Sandford ☎(0934) 852380
Holders
Weston Rd, Congresbury ☎(0934) 833034
Channel Service Station
30–36 Locking Rd, Weston-super-Mare ☎(0934) 20921
Victoria
Alfred St, Weston-super-Mare (BL DJ) ☎(0934) 21451
Moorland Motors
(Hutton Van Hire)
95 Moorland Rd, Weston-super-Mare ☎(0934) 27080

Oldmixon Service Station
(C H Cowie & Son),
Broadway, Weston-super-Mare (Peu Tal) ☎(0934) 812479
Reskes
Baker St, Weston-super-Mare (Cit) ☎(0934) 23995

Brent Knoll Rest Area
Northbound (THF)
☎(093472) 500
Fast food. Picnic area.
Shop. HGV parking.
Long-term/overnight parking for caravans £2.50. For disabled: toilets, ramp.
Credit cards – shop.

Junction 22

Burnham-on-Sea, a small resort on the Bristol Channel, is to the west.

WHERE TO STAY
★★*!♨***Woodlands**
Hill Ln, Brent Knoll ☎(0278) 760232
★**Battleborough Grange**
Bristol Rd, Brent Knoll ☎(0278) 760208
★★**Dunstan House**
8–10 Love Lane, Burnham-on-Sea ☎(0278) 784343
★★**Royal Clarence**
The Esplanade, Burnham-on-Sea (MINO) ☎(0278) 783138
★**Richmond**
32 Berrow Rd, Burnham-on-Sea ☎(0278) 782984
★★*!♨ HL* **Batch Farm**
Lympsham ☎(093472) 371
★★**Sundowner**
West Huntspill, Highbridge ☎(0278) 784766
FH Mrs E Puddy **Croft Farm**
Mark Causeway ☎(027864) 206

GARAGES
Brent Knoll
(I Steer) Brent Knoll (Lnc) ☎(0278) 760563
Autosteer
Church St, Highbridge ☎(0278) 783209
Leach's Accident Repair Centre
South St, Burnham-on-Sea ☎(0278) 789044
Youngs Car Centre
82 Oxford St, Burnham-on-Sea ☎(0278) 782323
Tarnock
Tarnock ☎(093472) 320

Behind the Memorial rises Brent Knoll, its summit capped by a hill-fort.

Junction 23

The distinctive landscape of Sedgemoor, cut across with a network of drainage channels, is to the east. At Bridgwater, a museum contains items from the Battle of Sedgemoor (1685).

WHERE TO EAT
××**Huntspill Villa**
82 Main Rd, West Huntspill ☎(0278) 782291
×**Wilton Farm House**
9 Goose Ln, Chilton Polden ☎(0278) 722134

GARAGES
Garage
Woolavington Rd, Puriton ☎(0278) 683280
Westway Cars
Taunton Rd, Bridgwater (Aud VW) ☎(0278) 426655
Harry Ball
Market St, Bridgwater (BL) ☎(0278) 422125
Bridgwater Motor Co
52 Eastover, Bridgwater (Ren) ☎(0278) 422218
K C Motor Services
Unit 19, Axe Rd, Colley Ln, Bridgwater ☎(0278) 426103
Motorcraft
Polden St, Bridgwater ☎(0278) 57240

Junction 24

To the west are the lovely Quantock Hills with heath and woodland, delightful villages and an abundance of wildlife.

WHERE TO STAY
★★**Walnut Tree Inn**
North Petherton (EXEC) ☎(0278) 662255
★★*L* **Royal Clarence**
Cornhill, Bridgwater ☎(0278) 55196
FH Mrs C M J Howard **Balls Farm** North Petherton ☎(0278) 662320

WHERE TO EAT
×**Old Vicarage**
45–49 St Mary St, Bridgwater ☎(0278) 58891

GARAGES
Heathfield
(Rawlinson) North Petherton ☎(0278) 662230

Junction 25

Taunton is an attractive market town on the River Tone. The Castle, scene of Judge Jeffreys' 'Bloody Assize' nearly 300 years ago, now contains the Somerset County Museum. Taunton is also the home of the Somerset County Cricket Club.

WHERE TO STAY
★★**St Quintin**
Bridgwater Rd, Bathpool ☎(0823) 73016
★★*H* **Falcon**
Henlade (EXEC) ☎(0823) 442502
★★**Creech Castle**
Bathpool ☎(0823) 73512
★★★**County**
East St, Taunton (THF) ☎(0823) 87651
✿★★★★*H* **Castle**
Castle Green, Taunton (PRE) ☎(0823) 72671

★★**Corner House**
Park St, Taunton ☎(0823) 84683
GH **White Lodge**
81 Bridgwater Rd, Taunton ☎(0823) 73287
GH **Ruishton Lodge**
Ruishton, Taunton ☎(0823) 442298
GH **Brookfield**
16 Wellington Rd, Taunton ☎(0823) 72786
GH **Meryan House**
Bishop's Hull Rd, Taunton ☎(0823) 87445

WHERE TO EAT
××**Quorum**
148 East Reach, Taunton ☎(0823) 88876
Crown Inn
Creech Heathfield ☎(0823) 412444

GARAGES
Wadham Stringer
Austin House, South St, Taunton (BL MG DJ LR Rar) ☎(0823) 88991
Rex Bros Motors
30–33a East Reach, Taunton (Tal) ☎(0823) 87871
Taunton
Bridgwater Rd, Bathpool (Ren Vlo Aud) ☎(0823) 412559
Taunton Auto Electrical and Diesel
Priorswood Rd, Taunton ☎(0823) 89111
Creech Motors
Creech St Michael ☎(0823) 442480
Co-operative Retail Services
Transport Dept
Magdelene St, Taunton ☎(0823) 75321
Thornfalcon Motors
Thornfalcon ☎(0823) 443065
Dunn Motors
43/45 East St, Taunton (Fia Cit) ☎(0823) 72607
South West Motor Services
Cornishway North, Galmington Trading Est, Taunton ☎(0823) 77805
Auto Safety Centre
Castle St, Taunton ☎(0823) 85691
Marshalsea Motors
Taunton Kawasaki Centre, 30–32 Wellington Rd, Taunton (Kaw Peu) ☎(0823) 76190

M5 Clevedon–Waterloo Cross 48 miles

After keeping close to the coastline of the Bristol Channel, the motorway turns inland with the Quantocks on one side and Sedgemoor on the other.

Taunton Deane
(Roadchef)
☎(0823) 71111
Restaurants. Picnic area.
Shop. Vending machines. Petrol. Diesel.
Breakdowns. Repairs.
HGV parking. Long-term/overnight parking for caravans £4 and HGV £5. Baby-changing. For disabled: toilets, ramp.
Footbridge. Credit cards – garage. Tourist information services.

Flat marshy Sedgemoor is drained by a network of 'rhines' or ditches.

Junction 26

Wellington, the town which lent its name to the Iron Duke, is to the north. South of the town, and on the opposite side of the motorway, is the 175ft high Wellington Monument.

WHERE TO STAY
★★**Heatherton Grange**
Wellington Rd, Bradford-on-Tone ☎(0823) 46777
★★**Beam Bridge**
Sampford Arundel ☎(0823) 672223
GH **Rumwell Hall**
Rumwell, Taunton ☎(0823) 75268

WHERE TO EAT
××**Well House**
Poundisford ☎(082342) 566

GARAGES
M D Motorcycles
16–18 North St, Wellington (Gar) ☎(082347) 4076
Richardsons
44 High St, Wellington (BL) ☎(082347) 4181

Junction 27

At Uffculme, to the east, is Coldharbour Mill, an 18th-century working cloth mill producing knitting wool and woven cloth. Tiverton has a historic castle and an interesting folk museum.

WHERE TO STAY
★**Green Headland**
Sampford Peverell ☎(0884) 820255
FH Mrs H M Parkhouse
Higher Shutehanger Farm
Sampford Peverell ☎(0884) 820569
FH Mrs M D Farley
Houndaller Farm Uffculme ☎(0884) 40246
FH Mrs C M Baker **Woodrow Farm** Uffculme ☎(0884) 40362
FH Mrs J M Granger **Doctors Farm** Willand ☎(0884) 820525

WHERE TO EAT
Farm House Inn
Leonard's Moor, Sampford Peverell ☎(0884) 820824
Poachers Pocket
Burlescombe, Wellington ☎(0823) 672286

GARAGES
Station
Bridge St, Uffculme ☎(0884) 40247
Willand
(L J Spearing & Son) Willand ☎(0884) 820224
Lamb Hill
(A38) Burlescombe ☎(0884) 40664
Culmstock
Culmstock ☎(0884) 40570

LOCAL RADIO STATIONS

| | Medium Wave | | VHF/FM |
	Metres	kHz	MHz
BBC Radio Bristol	194	1548	95.5
IBA Radio West	238	1260	96.3

M5 Waterloo Cross – Exeter   19 miles

The motorway follows the River Culm through lovely Devon countryside before passing east of Exeter.

Junction 27 – see page 63

Junction 28

The small market town of Cullompton is beside this junction. To the west a network of narrow lanes crosses steep hilly farmland.

WHERE TO STAY
FH Mrs A C Cole **Five Bridges Farm** Cullompton
☎(0884) 33453
FH Mrs J M Granger **Doctors Farm** Willand
☎(0884) 820525
FH Mrs B J Hill **Sunnyside Farm** Butterleigh
☎(08845) 322

GARAGES
Highway
Unit 2, Whittons Transport Yd, Station Rd, Cullompton
☎(0884) 33316
Clark's Motors
High St, Cullompton (BL)
☎(0884) 32345
Culm
1 Willand Rd, Cullompton (Frd) ☎(0884) 33551

Junction 29 R

No exit for southbound traffic here. Northbound travellers will find Exeter's local airport to the east at Clyst Honiton, while the road west leads into the city.

WHERE TO EAT
PS **Clyst Honiton**
¾m E of Clyst Honiton on A30. OS192 SX9994

GARAGES
Rockbeare Motor Services
London Rd, Rockbeare
☎(0404) 410

Junction 30

Exeter has a long history and much to interest present day visitors. Historic buildings include the Cathedral, Guildhall and St Nicholas' Priory. Of the city's museums the Maritime Museum is particularly noteworthy.

Exeter Services
(Granada) ☎(0392) 36267
Restaurant. Fast food. Shop. Playground. Petrol. Diesel. Breakdowns. Repairs. HGV parking. Long-term/overnight parking for caravans £3.50 per night. Baby-changing. For disabled: toilets, ramp. Credit cards – shop, garage, restaurant.

WHERE TO STAY
☆☆☆ **Exeter Moat House**
Topsham Rd, Exeter Bypass (QM) ☎(039287) 5441

☆☆☆ **Devon Motel**
Exeter Bypass, Matford (PRE)
☎(0392) 59268
★★★ **Gipsy Hill**
Pinhoe, Exeter
☎(0392) 65252
★★★ **Buckerell Lodge Crest**
Topsham Rd, Exeter (CRH)
☎(0392) 52451
GH **Willowdene**
161 Magdelen Rd, Exeter
☎(0392) 71925
GH **Westholme**
85 Heavitree Rd, Exeter
☎(0392) 71878
GH **Regents Park**
Polsloe Rd, Exeter
☎(0392) 59749
FH Mrs A Freemantle
Ivington Farm Clyst St Mary
☎(039287) 3290

WHERE TO EAT
Amadeus Restaurant
62 Fore St, Topsham
☎(039287) 3759
Clare's
13 Princesshay, Exeter
☎(0392) 55155
Coolings Wine Bar
11 Gandy St, Exeter
☎(0392) 34183
Ship Inn
Martin's Lane, Exeter
☎(0392) 72040

GARAGES
Sports Classic & Performance
No 6 Cygnets Units, Heron Rd, Sowton Industrial Est, Exeter ☎(0392) 59786
C & P Motors
Clyst St Mary (Ren)
☎(039287) 5000

Standfield & White
Honiton Rd, Exeter (Dat)
☎(0392) 68187
Mote Service Station
85 Fore St, Heavitree, Exeter (Lnc) ☎(0392) 56790
Seabrook Service Station
Topsham Rd, Countess Weir, Exeter ☎(039287) 7272
High St
(W W Pretty & Son) Topsham
☎(039287) 3056
Ansons
Exhibition Way, Pinhoe Trading Est, Exeter (Fia)
☎(0392) 69352
Countess Weir Self Serve
399 Topsham Rd, Exeter
☎(039287) 3110
Harrison Brett
210 Monks Rd, Exeter
☎(0392) 74786
Honiton Clyst
Clyst Honiton, Exeter
☎(0392) 67235

Junction 31

The end of the motorway and the gateway to Dartmoor and the Torbay resorts of Torquay, Paignton and Brixham.

WHERE TO STAY
★★ **Fairwinds**
Kennford ☎(0392) 832911
☆☆ **Ladbroke**
(on A38) Kennford (LB)
☎(0392) 832121
★★♨ **Trood Country House**
Little Silver Ln, Alphington, Exeter ☎(0392) 75839
GH **Gledhills**
32 Alphington Rd, Exeter
☎(0392) 71439
FH Mrs R Weeks **Holloway Barton Farm** Kennford
☎(0392) 832302

Exeter's Maritime Museum has the biggest boat collection of is kind in the world. Pictured above is Butt's Ferry.

WHERE TO EAT
Swan's Nest
Exminster, Exeter
☎(0392) 832371
Haldon Thatch
Bottom of Telegraph Hill, Kennford ☎(0392) 832273

GARAGES
Kennford
Main Rd, Kennford
☎(0392) 832050
Carburation Services
8 Church Rd, Alphington, Exeter ☎(0392) 73772
Woods Western
Alphington Rd, Exeter
☎(0392) 74458
G Tancocks
Landscore Rd, St Thomas, Exeter ☎(0392) 55686

LOCAL RADIO STATIONS			
	Medium Wave		**VHF/FM**
	Metres	*kHz*	*MHz*
BBC Radio Devon	375	801	96.2
Exeter area	351	855	97.5
IBA DevonAir Radio			
Exeter area	450	666	95.8

M6 Rugby – Coleshill   23 miles

At the heart of the country's motorway system the M6 branches off from the M1, heading for all points north-west.

Junction 1

Rugby is to the south (see also M1 junction 18, page 36). The industry which grew in the past around the canal and railway on the north side of town is now easily accessible from this motorway.

WHERE TO STAY
★★★♨ **Clifton Court**
Lilbourne Rd, Clifton upon Dunsmore, Rugby
☎(0788) 65033
★★★ **Three Horse Shoes**
Sheep St, Rugby (PG)
☎(0788) 4585
★★ **Moorbarns**
Watling St, Lutterworth
☎(04555) 2237
★★ **Hillmorton Manor**
High St, Hillmorton, Rugby
☎(0788) 65533
GH **Grosvenor House**
81 Clifton Rd, Rugby
☎(0788) 3437
GH **Mound**
17–19 Lawford Rd, Rugby
☎(0788) 3486

WHERE TO EAT
✕**Andalucia**
10 Henry St, Rugby
☎(0788) 76404
Carlton
Railway Ter, Rugby
☎(0788) 3076

GARAGES
Gibbetts Cross
Gibbetts Cross, Watling St, Churchover ☎(0788) 860215
Parkside
Railway Ter, Rugby (BL)
☎(0788) 3477
Ron Forster Cars
Leicester Rd, Rugby (Hon)
☎(0788) 2934
Grove
Forum Dr, Leicester Rd, Rugby (Tal Peu)
☎(0788) 62731
Rugby Ignition and Carburettor Service
Unit 5C, 21 Consul Rd, Rugby
☎(0788) 78082
Hillmorton
102 Hillmorton Rd, The Parade, Hillmorton, Rugby
☎(0788) 2515

Junction 2

The M69 heads north to Leicester (see page 84 for details) while the road south leads to the centre of Coventry, an industrial city with a famous post-war cathedral. The Museum of British Road Transport illustrates Coventry's contribution to this industry over the years. Other places of interest include a toy museum and a medieval guildhall.

WHERE TO STAY
☆☆☆ **Crest**
Hinckley Rd, Walsgrave, Coventry (CRH)
☎(0203) 613261
GH **Croft**
23 Stoke Gn, off Binley Rd, Coventry ☎(0203) 457846

WHERE TO EAT
✕✕**Grandstand**
Coventry City F.C., King Richard St, Highfield, Coventry ☎(0203) 27053

GARAGES
Potters Green Service Station
Ringwood Highway, Potters Green, Coventry
☎(0203) 614570
International Auto Safety Centre
219 Swan Ln, Walsgrove, Coventry ☎(0203) 21574
Pippin
167 Ansty Rd, Wyken, Coventry ☎(0203) 444300
Binley Woods Service Station
60–62 Rugby Rd, Binley Woods, Coventry
☎(0203) 542202
BWL Motors
450 Stoney Stanton Rd, Coventry ☎(0203) 85019

Junction 3

This junction separates Coventry's northern outskirts from the town of Bedworth and its expanding industrial estate.

WHERE TO STAY
☆☆☆ **Novotel Coventry**
Wilsons Ln, (A444), Coventry
☎(0203) 365000
★★★ **Royal Court**
Tamworth Rd, Keresley, Coventry ☎(020333) 4171
★★ **Beechwood**
Sandpits Ln, Keresley, Coventry ☎(020333) 4243
★★ **Griff House**
Coventry Rd, Nuneaton
☎(0203) 382984

WHERE TO EAT
Corks Wine Bar
Whitefriars St, Coventry
☎(0203) 23628
Nello Pizzeria
8 City Arcade, Coventry
☎(0203) 23551

GARAGES
Mann Egerton
Holbrooks Ln, Coventry (Peu Tal) ☎(0203) 83121
Parkside
Lockhurst Ln, Coventry (BL)
☎(0203) 88851
City
Roseberry Av, Bell Green, Coventry (Mos)
☎(0203) 87904

Junction 4

The M42 leads south to the National Exhibition Centre and Birmingham Airport – see page 83 for details.

Corley Services
(THF) ☎(0676) 40111
24 hr cafeteria. "Little Chef". Waitress service (north side)
07.30–22.00. Fast food. Picnic area. Shop. Playground. Petrol. Diesel. Breakdowns. Repairs. HGV parking. Long-term/overnight parking for caravans £2.50. Baby-changing. For disabled: toilets, ramp. Footbridge. Credit cards – shop, garage, catering.

WHERE TO STAY
★★ **Swan**
High St, Coleshill (AB)
☎(0675) 62212
★★★★ **Excelsior**
Coventry Rd, Elmdon, Birmingham Airport (THF)
☎021-743 8141
GH **Heath Lodge**
Coleshill Rd, Marston Green, Birmingham ☎021-779 2218
GH **Tri-Star**
Coventry Rd, Elmdon, Birmingham ☎021-779 2233
INN **George & Dragon**
154 Coventry Rd, Coleshill
☎(0675) 62249

GARAGES
Chelmsley Wood
Chester Rd, Bacons End, Chelmsley Wood, Birmingham (Ren)
☎021-770-8373
Skyways Service Station
Coventry Rd, Birmingham Airport ☎021-742 8943
Hillcrest
Grimstock Hill, Lichfield Rd, Coleshill ☎(0675) 62190

LOCAL RADIO STATIONS			
	Medium Wave		**VHF/FM**
	Metres	*kHz*	*MHz*
BBC Radio Leicester	358	837	95.1
BBC Radio WM	206	1458	95.6
IBA Mercia Sound	220	1359	95.9

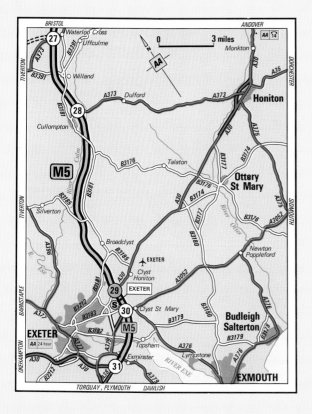

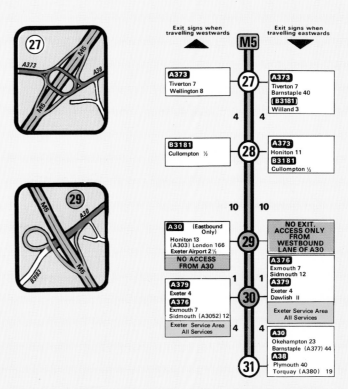

Exit signs when travelling westwards		Exit signs when travelling eastwards
	M5	
A373 Tiverton 7 Wellington 8	**27**	**A373** Tiverton 7 Barnstaple 40 **B3181** Willand 3
	4 4	
B3181 Cullompton ½	**28**	**A373** Honiton 11 **B3181** Cullompton ½
	10 10	
A30 (Eastbound Only) Honiton 13 (A303) London 166 Exeter Airport 2½ **NO ACCESS FROM A30**	**29**	**NO EXIT. ACCESS ONLY FROM WESTBOUND LANE OF A30**
	1 1	**A376** Exmouth 7 Sidmouth 12 **A379** Exeter 4 Dawlish II
A379 Exeter 4 **A376** Exmouth 7 Sidmouth (A3052) 12 Exeter Service Area All Services	**30**	Exeter Service Area All Services
	4 4	**A30** Okehampton 23 Barnstaple (A377) 44 **A38** Plymouth 40 Torquay (A380) 19
	31	

M6 **Rugby - Coleshill**

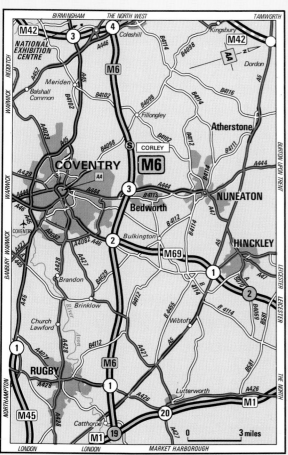

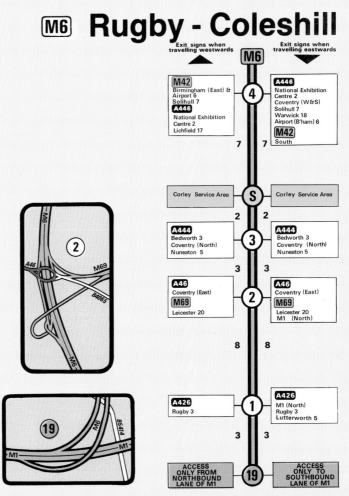

Exit signs when travelling westwards		Exit signs when travelling eastwards
	M6	
M42 Birmingham (East) & Airport 6 Solihull 7 **A446** National Exhibition Centre 2 Lichfield 17	**4**	**A446** National Exhibition Centre 2 Coventry (W & S) Solihull 7 Warwick 18 Airport (B'ham) 6 **M42** South
	7 7	
Corley Service Area	**S**	Corley Service Area
	2 2	
A444 Bedworth 3 Coventry (North) Nuneaton 5	**3**	**A444** Bedworth 3 Coventry (North) Nuneaton 5
	3 3	
A46 Coventry (East) **M69** Leicester 20	**2**	**A46** Coventry (East) **M69** Leicester 20 M1 (North)
	8 8	
A426 Rugby 3	**1**	**A426** M1 (North) Rugby 3 Lutterworth 5
	3 3	
ACCESS ONLY FROM NORTHBOUND LANE OF M1	**19**	**ACCESS ONLY TO SOUTHBOUND LANE OF M1**

Coleshill - Keele M6

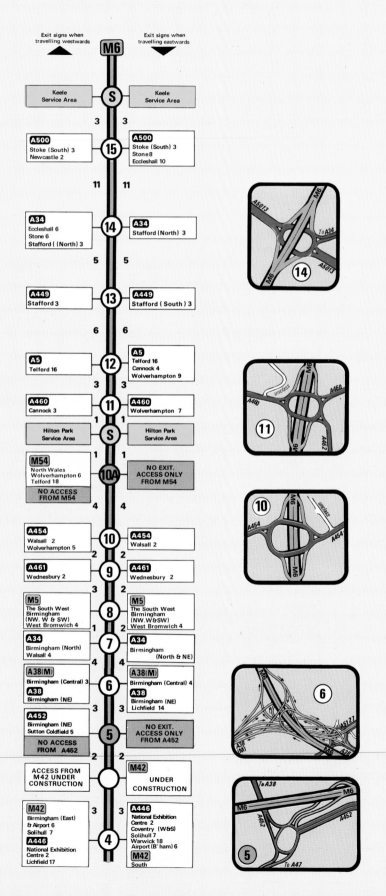

Exit signs when travelling westwards ◄   Exit signs when travelling eastwards ▼

M6

Keele Service Area — S — **Keele Service Area**

3   3

A500 Stoke (South) 3 / Newcastle 2 — 15 — **A500** Stoke (South) 3 / Stone 8 / Eccleshall 10

11   11

A34 Eccleshall 6 / Stone 6 / Stafford (North) 3 — 14 — **A34** Stafford (North) 3

5   5

A449 Stafford 3 — 13 — **A449** Stafford (South) 3

6   6

A5 Telford 16 — 12 — **A5** Telford 16 / Cannock 4 / Wolverhampton 9

3   3

A460 Cannock 3 — 11 — **A460** Wolverhampton 7

1   1

Hilton Park Service Area — S — **Hilton Park Service Area**

1   1

M54 North Wales / Wolverhampton 6 / Telford 18 / NO ACCESS FROM M54 — 10A — NO EXIT. ACCESS ONLY FROM M54

4   4

A454 Walsall 2 / Wolverhampton 5 — 10 — **A454** Walsall 2

2   2

A461 Wednesbury 2 — 9 — **A461** Wednesbury 2

3   3

M5 The South West / Birmingham (NW. W & SW) / West Bromwich 4 — 8 — **M5** The South West / Birmingham (NW. W & SW) / West Bromwich 4

1   2   2   1

A34 Birmingham (North) / Walsall 4 — 7 — **A34** Birmingham (North & NE)

4   4

A38(M) Birmingham (Central) 3 / **A38** Birmingham (NE) — 6 — **A38(M)** Birmingham (Central) 4 / **A38** Birmingham (NE) / Lichfield 14

3   3

A452 Birmingham (NE) / Sutton Coldfield 5 / NO ACCESS FROM A452 — 5 — NO EXIT. ACCESS ONLY FROM A452

2   2

ACCESS FROM M42 UNDER CONSTRUCTION — ◯ — **M42** UNDER CONSTRUCTION

3   3

M42 Birmingham (East) & Airport 6 / Solihull 7 / **A446** National Exhibition Centre 2 / Lichfield 17 — 4 — **A446** National Exhibition Centre 2 / Coventry (W&S) / Solihull 7 / Warwick 18 / Airport (B'ham) 6 / **M42** South

14   11   10   6   5

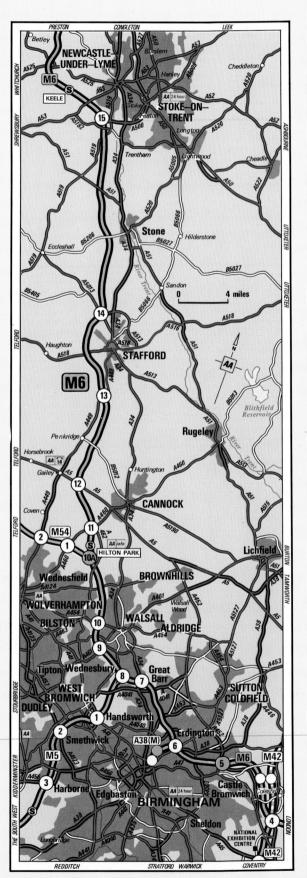

Junction 4 – see page 64

Junction 5 R

A restricted junction, but northbound travellers can leave here for Castle Bromwich and the north-eastern suburbs of Birmingham.

WHERE TO STAY
★★★★**Penns Hall**
Penns Ln, Walmley, Sutton Coldfield (EH)
☎021-351 3111
GH **Cape Race**
929 Chester Rd, Erdington
☎021-373 3085
GH **Hurstwood**
775–777 Chester Rd, Erdington ☎021-382-8212

GARAGES
Albany
Murco Service Station
Turnhouse Rd, Castle Vale, Castle Bromwich
☎021-747 6326
Kings
25–27 Coleshill Rd, Ward End (BMW Hon Suz Yam)
☎021-784 0833
Charters
1 School Ln, Kitts Green
☎021-783 2416

Junction 6

The notorious 'spaghetti junction' where a network of slip-roads pass over and under the main motorway to link with the Aston Expressway into central Birmingham.

WHERE TO STAY
★★★**Royal Angus Thistle**
St Chads, Queensway, Birmingham (TS)
☎021-236 4211
★★**New Imperial**
Temple St, Birmingham
☎021-643 6751
★★★**Grand**
Colmore Row, Birmingham (QM) ☎021-236 7951
★★★★**Albany**
Smallbrook, Queensway, Birmingham (THF)
☎021-643 8171
GH **Lyndhurst**
135 Kingsbury Rd, Erdington
☎021-373 5695
GH **Rollason Wood**
130 Wood End Rd, Erdington
☎021-373 1230
GH **Lomax House**
36 Trinity Rd, Birchfield
☎021-554 3951
GH **Linden Lodge**
79 Sutton Rd, Erdington
☎021-382 5992

WHERE TO EAT
××**Lorenzo's**
Park St, Birmingham
☎021-643 0541
✸××**Rajdoot**
12–22 Albert St, Birmingham
☎021-643 8805
××**New World**
308 Bull Ring Centre, Small Brook Ringway, Birmingham
☎021-643 0033
××**New Happy Gathering**
43–45 Station St, Birmingham ☎021-643 5247

GARAGES
Bristol St Motors
Beacon House, Long Acre, Nechells (Frd)
☎021-327 4791
Midland Link Service Station
288–298 Tyburn Rd, Erdington ☎021-350 4493

Bannings Car Safety Centre
501/523 Lichfield Rd, Aston
☎021-327 2448
Colmore Car People
Witton Ln, Aston (Fia Lnc Fer)
☎021-328 7777
Whites
257 Great Lister St, Birmingham ☎021-359 3571
Colmore Car People
Sutton New Rd, Erdington
(Fia Lnc Dai) ☎021-350 1301
Yenton Motor Co
103 Goosemore Ln, Erdington (Tal Peu)
☎021-382 1919
Colliers
(Abbey) 37 Sutton Rd, Erdington (BL)
☎021-382 5000
A F Jameson
27–30 Summer Ln, Newtown, Aston ☎021-359 7991
Kings
25–27 Coleshill Rd, Ward End (BMW Hon Suz Yam)
☎021-784 0833
Midlands Motor Cycles
(Forumburn Ltd) 529–531 Coventry Rd, Small Heath (Hon Suz Yam Kaw)
☎021-772 1733
George Heath Motors
(Talbot Motor Co) 90–94 Charlotte St, Birmingham (Tal Peu) ☎021-236 4382

Junction 7

There is a viewpoint at Barr Beacon to the north, while eastwards is Sutton Park, a lovely area of lakes and woodland.

WHERE TO STAY
☆☆☆**Post House**
Chapel Ln, Great Barr (THF)
☎021-357 7444
★★★**Barr**
Pear Tree Dr, off Newton Rd, Great Barr (GW)
☎021-357 1141
☆☆**Parson and Clerk Motel**
Chester Rd, Streetly (AB)
☎021-353 1747

WHERE TO EAT
PS **Sandwell Valley**
Park Lane, West Bromwich
OS139 SP0292
PS **Sutton Park**
1m W of centre of Sutton Coldfield OS139 SP1096

GARAGES
Thame Service Station
(D A Mew) 230 Birmingham Rd, Great Barr
☎021-357 7960
Station Drive
Thornhill Rd, Sutton Coldfield
☎021-353 7322

Junction 8

The M5 heads southwards from this junction – see pages 58–64 for details.

Junction 9

At Wednesbury is the Sandwell Art Gallery and Museum which includes a collection of 19th-century English paintings, applied arts and local history displays.

WHERE TO STAY
☆☆☆**Crest**
Birmingham Rd, Walsall (CRH) ☎(0922) 33555
★★**County**
Birmingham Rd, Walsall (QM)
☎(0922) 32323

GARAGES
Clark Walsall
Bull Stake, Wednesbury Rd, Darlaston (BL DJ)
☎021-526 2263
Workshop
58–60 Station St, Darlaston
☎021-526 3987
Rapid Auto Body Repairers
Bridge St, Wednesbury
☎021-556 0549

Junction 10

Walsall, once famous for its leather work, is now the home of more diverse industries, but a Museum of Leathercraft is included in the Walsall Museum and Art Gallery. Nearby Willenhall has a museum which traces its lock-making history.

WHERE TO STAY
★★**Royal**
Ablewell St, Walsall
☎(0922) 24555

GARAGES
Walsall Way Service Station
Wolverhampton Rd, Walsall
☎(0922) 25372
Primley Service Station
Wolverhampton Rd, Walsall
☎(0922) 34532
Hartshorne Motor Services
Bentley Mill Cl, Walsall
☎(0922) 20941
Hewitts
Wolverhampton St, Walsall
(BL DJ LR Rar)
☎(0922) 26567
County Bridge Service Station
Wolverhampton Rd West, Willenhall ☎(0902) 65734
Bridge St Service Station
Bridge St, Walsall
☎(0922) 35525
Falcon
(J Haycock & Sons) Rough Hay Rd, Darlaston
☎021-526 2168

Junction 10A R

The M54 branches off westwards here heading for Telford – see page 84 for details.

Hilton Park Services
(Rank) ☎(0922) 415537
Restaurant. Fast food. HGV cafe. Picnic area. Shop. Petrol. Diesel. Liquid Petroleum Gas. Breakdowns. Repairs. HGV parking. Long-term/overnight parking for caravans £3. (max 24hrs). Baby-changing. For disabled: toilets, lift. Footbridge. Roadbridge. Credit cards accepted.

Junction 11

Cannock is a small industrial town on the edge of the beautiful Cannock Chase, which is about 25 square miles of moors and woodland.

WHERE TO STAY
★★**Hollies**
Hollies Av, Cannock
☎(05435) 3151

GARAGES
Hilton Service Station
Cannock Rd, Featherstone
☎(0902) 732566

M6 Coleshill – Keele   52 miles

For the first part the motorway winds through the built-up areas of the West Midlands connurbation. Later it crosses more attractive Staffordshire countryside, heading for the Potteries.

Birmingham's Gas Street Basin is one of our best known canal areas.

Churchbridge Motor
Watling St, Cannock (Frd)
☎(0922) 417014
Gold Star
Watling St, Cannock (Vau Opl) ☎(05435) 5361

Junction 12

The A5, which links with the motorway here, is the old Roman Watling Street. Penkridge, to the north, has a particularly interesting church, said to be the finest in Staffordshire.

WHERE TO STAY
☆☆☆**Roman Way**
Watling St, Hatherton (EXEC)
☎(05435) 72121

GARAGES
Gailey Service Station
Watling St (A5), Gailey
☎(0902) 790589
Croft
(E W & P Barrow) Brewood Rd, Coven (Sub)
☎(0902) 790217

Junction 13

The motorway runs parallel to both the River Penk and the Staffordshire and Worcestershire Canal here. The county town of Stafford is to the north and Cannock Chase Country Park to the east.

WHERE TO STAY
★**Garth**
Wolverhampton Rd, Moss Pit, Stafford (WDB)
☎(0785) 56124
★**Romaline**
73 Wolverhampton Rd, Stafford ☎(0785) 54100
GH **Abbey**
65–68 Lichfield Rd, Stafford
☎(0785) 58531
GH **Leonards Croft**
80 Lichfield Rd, Stafford
☎(0785) 3676

WHERE TO EAT
PS **Cannock Chase**
Milford Village
OS139 SJ9721

GARAGES
Brocton Service Station
Cannock Rd, Brocton, Stafford ☎(0785) 661311
Dunns
Stafford Rd, Penkridge
☎(078571) 2260
Rising Brook Service Station
Wolverhampton Rd, Stafford
☎(0785) 58242

Junction 14

This junction gives access to the north of Stafford. At Shallowford, to the north-west, is Izaak Walton's Cottage where the famous angler lived.

WHERE TO STAY
☆☆☆**Tillington Hall**
Eccleshall Rd, Stafford (GW)
☎(0785) 53531
★★**Swan**
Greengate St, Stafford (Bl)
☎(0785) 58142
★**Vine**
Salter St, Stafford (WDB)
☎(0785) 51071

WHERE TO EAT
×**Holly Bush Inn**
Seighford ☎(078575) 280
××**Worston Mill**
Worston, Great Bridgeford
☎(078575) 710
×××**Yew Tree**
Ranton ☎(078575) 278
Annabel's
Greyfriars, Stafford
☎(0785) 54500
Anemos
22 Crabbery St, Stafford
☎(0785) 48940

GARAGES
Lloyds
Stone Rd, Stafford (Frd)
☎(0785) 51331
Sandon Rd Motors
Sandon Rd, Stafford
☎(0785) 45219
Station
(Fred Shaw & Son) Derby St, Stafford (Maz)
☎(0785) 55486
Queensville
(Charles Clarke & Son)
Lichfield St, Stafford
☎(0785) 51366
Lammascote Service Station
Lammascote Rd, Stafford
☎(0785) 51107

Junction 15

Here in The Potteries are a number of places to see examples of Staffordshire pottery and porcelain – being made and on display: Wedgwood,

---

Spode, Gladstone Pottery and the Sir Henry Doulton Gallery. Stoke-on-Trent's City Museum and Art Gallery has a particularly fine collection. Close to the junction are Trentham Gardens which cover 700 acres.

WHERE TO STAY
★★★**Clayton Lodge**
Newcastle Rd, Clayton (EH)
☎(0782) 613093
☆☆**Post House**
Clayton Rd, Newcastle-under-Lyme (THF)
☎(0782) 625151
★**H Deansfield**
98 Lancaster Rd, Newcastle-under-Lyme ☎(0782) 619040
★★**Borough Arms**
Kings St, Newcastle-under-Lyme ☎(0782) 629421
GH **Grove Court**
100 Lancaster Rd, Newcastle-under-Lyme
☎(0792) 614406

WHERE TO EAT
××**Poacher's Cottage**
Stone Rd, Trentham, Stoke-on-Trent ☎(0782) 657115
××**Cock Inn**
Stableford ☎(0782) 680210
Capri Ristorante Italiano
13 Glebe St, Stoke-on-Trent
☎(0782) 411889
PS **Hanchurch Hills**
3m S of Newcastle-under-Lyme OS127 SJ8439

GARAGES
Swift Service Station
(M & S Auto Care) Clayton Rd, Clayton ☎(0782) 636446
Hanford Motor Services
Hanford Rbt, Stoke-on-Trent
☎(0782) 657037
Highland
Higherland, Newcastle-under-Lyme (Vau Opl)
☎(0782) 610941
Newcastle Motors
Hassell St, Newcastle-under-Lyme (Peu Tal Cit Col)
☎(0782) 614621
Rix Manor
Brook Ln, Newcastle-under-Lyme (BL) ☎(0782) 618461

Keele Services
(THF) ☎(0782) 626221
Cafeteria 07.30–22.00. Waitress service 'Little Chef' 07.00–22.00. Fast food. Shop. Petrol. Diesel. Breakdowns. Repairs. HGV parking. Long-term/overnight parking for caravans £2.50. Baby-changing. For disabled: toilets, ramp. Footbridge. Credit cards – shop, garage, restaurants.

LOCAL RADIO STATIONS

	Medium Wave		VHF/FM
	Metres	kHz	MHz
BBC Radio WM	206	1458	95.6
BBC Radio Stoke-on-Trent	200	1503	94.6
IBA BRMB	261	1152	94.8
IBA Beacon Radio	303	990	97.2
IBA Signal Radio	257	1170	104.3

M6 Keele – Charnock Richard   53 miles

The motorway continues northwards through the Cheshire countryside before entering the industrial areas of Merseyside and southern Lancashire.

**Keele Services –
see page 67**

The one-star Rose and Crown in Knutsford was a coaching inn, built in 1647. Although it was restored in 1923, it retains its original style.

Junction 16

On the north side of Stoke-on-Trent is Ford Green Hall, a timber-framed yeoman farmer's house, and the Chatterley Whitfield Mining Museum with guided tours of the mine workings.

GARAGES
Wrinehill
Newcastle Rd, Wrinehill
☎(0270) 820300

Sandbach Services
(Roadchef)
☎(09367) 7134
Restaurant. Fast food. Shop. Tourist Information. Vending machines. Petrol. Diesel. Breakdowns. Repairs. HGV parking. Long-term/overnight parking for caravans £5. Baby-changing. For disabled: toilets, ramp. Footbridge. Credit cards – garage.

Junction 17

The historic town of Sandbach is famous for its two ancient Saxon crosses which stand in the cobbled Market Square.

WHERE TO STAY
★★★**Chimney House**
Congleton Rd, Sandbach
(WWP) ☎(09367) 4141
☆☆**Saxon Cross Motel**
Holmes Chapel Rd, Sandbach ☎(09367) 3281
★★**Old Hall**
Newcastle Rd, Sandbach
☎(09367) 61221

GARAGES
A & D Beech Motor Eng
rear of Military Arms Hotel, Congleton Rd, Sandbach
☎(0270) 585193

Sandbach Service Station
Bradwell Rd, Sandbach
☎(09367) 3395
Old Smithy
(Pace Arclid) Arclid
☎(04775) 334
Ettiley Heath
(L & S Motors) Elston Rd, Sandbach ☎(09367) 60625
Station
(G C & J F Makinson) Moss Ln, Elworth
☎(09367) 2900

Junction 18

The Trent and Mersey Canal, Shropshire Union Canal and the Rivers Dane and Weaver all pass through Middlewich, a major salt–producing town.

WHERE TO EAT
××**Yellow Broom**
Twemlow Green, Holmes Chapel ☎(0477) 33289

GARAGES
R Gibson Vehicles
London Rd, Holmes Chapel
☎(0477) 32103
B S Morgan & Son
Main Rd, Goostrey

Knutsford Services
(Rank)
☎(0565) 4138 (garage)
Restaurant. Fast food. HGV cafe. Picnic area. Shop. Petrol. Diesel. Breakdowns. Repairs. HGV parking. Long-term/overnight parking for caravans £3 (max 24hrs). For disabled: toilets, lift. Footbridge. Road bridge. Credit cards accepted.

Junction 19

The charming old town of Knutsford is to the east.

★**Lymm**
Whitbarrow Rd, Lymm (GW)
☎(092575) 2233

GARAGES
Howarth Motors
101 Knutsford Rd, Grappenhall (VW Aud)
☎(0925) 65265
Vernon Motors
Black Bear Bridge, Knutsford Rd, Latchford
☎(0925) 35326
Park
(Avenue Motor Services)
Agden ☎(092575) 2447
Station
47 Mill Ln, Heatley
☎(092575) 2231
Ring O'Bells Service Station
Northwich Rd, Lower Stretton
☎(092573) 551
International Auto & Safety Centre
Priestley St, Warrington
☎(0925) 33308

Junction 21

The town of Warrington, standing on the River Mersey and the Manchester Ship Canal, was once the 'beer capital of Britain', but is better known these days for its vodka.

WHERE TO STAY
★★**Patten Arms**
Parker St, Warrington

GARAGES
Kingsway Service Station
Kingsway North, Warrington
☎(0925) 822205

Junction 21A

The M62 crosses here – see pages 86–93 for details.

Junction 22

Travellers from the north will leave here for Warrington, which may not be the most attractive of our towns, but a town trail points out its more interesting buildings.

GARAGES
Smithy
(Essendy) 32 Church Ln, Culcheth ☎(092576) 4419
International Auto & Safety Centre
Priestley St, Warrington
☎(0925) 33308

Junction 23

Haydock Park racecourse is right beside this junction.

WHERE TO STAY
☆☆☆☆**Post House**
Lodge Ln, Haydock (THF)
☎(0942) 717878

GARAGES
A B Motors
Mill Ln, Newton-le-Willows
(Dat) ☎(09252) 4411

Junction 24 R

The industrial and coal-mining town of Ashton in Makerfield is to the east.

GARAGES
Bickershaw Lane
12 Bickershaw Ln, Abram
☎(0924) 866418

Junction 25 R

A branch of the motorway leads towards Wigan from this junction for northbound travellers. It is a busy manufacturing town on the canal where the famous Wigan pier can be found.

Junction 26

The M58 leads to Aintree from this junction – see page 87 for details. To the east is Wigan.

WHERE TO STAY
★★**Holland Hall**
6 Lafford Ln, Up Holland
☎(0695) 624426
★★★**Brocket Arms**
Mesnes Rd, Wigan
☎(0942) 46283
INN **Coach House**
240A Warrington Rd, Lower Ince ☎(0942) 866330

WHERE TO EAT
Roberto's
Rowbottom Sq, Wigan
☎(0942) 42385

GARAGES
Delph Service Station
Billinge Rd, Pemberton
☎(0942) 214312
Gordon Ford – Wigan
Wallgate, Wigan (Frd)
☎(0942) 41393
Pepper Mill Exhausts
Pepper Mill, Darlington St, Wigan ☎(0942) 492025
Wilsons
270 Manchester Rd, Higher Ince ☎(0942) 492088

Junction 27

The town of Standish is to the east, but if you travel westwards you will find a viewpoint which looks out across the river and canal to the south.

WHERE TO STAY
☆☆☆**Cassinelli's Almond Brook Motor Inn**
Almond Brook Rd, Standish
☎(0257) 425588
★★**Lindley**
Lancaster Ln, Parbold
☎(02576) 2804
★★**Bel-Air**
236 Wigan Ln, Wigan
☎(0942) 41410
★★**Bellingham**
149 Wigan Ln, Wigan
☎(0942) 43893

WHERE TO EAT
Bollin Restaurant
Heatley ☎(092575) 3657

GARAGES
Standish Service Station
Preston Rd, Standish
☎(0257) 422899
Foxfield
(J Berry) Wigan Rd, Shevington (BL)
☎(02575) 2626
Gibsons Car Care
275 Mossylea Rd, Wrightington
☎(0257) 421668

Charnock Richard Services
(THF)
☎(0257) 791494
Accommodation. 24hr cafeteria. Waitress-service 'Little Chef' 07.30–22.00. Shop. Petrol. Diesel. Breakdowns. Repairs. HGV parking. Long-term/overnight parking for caravans £2.50. Baby-changing. For disabled: toilets, ramp. Credit cards – shop, garage, hotel, restaurants.

Jodrell Bank has one of the largest fully steerable radio telescopes in the world.

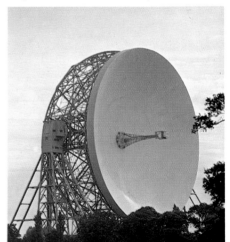

LOCAL RADIO STATIONS

	Medium Wave		VHF/FM
	Metres	kHz	MHz
BBC Radio Stoke-on-Trent	200	1503	94.6
BBC Radio Manchester	206	1458	95.1
BBC Radio Merseyside	202	1458	95.8
BBC Radio Lancashire	351	855	96.4
IBA Signal Radio	257	1170	104.3
IBA Piccadilly Radio	261	1152	97.0
IBA Radio City	194	1548	96.7

Junction 20 R

Lymm is an attractive small town, at the centre of which is an old market cross and stocks on a cobbled square. The M56 can be reached from this junction – see pages 86–89 for details.

WHERE TO STAY
★**Rockfield**
3 Alexandra Rd, Grappenhall
☎(0925) 62898
★**Ribblesdale**
Balmoral Rd, Grappenhall
☎(0925) 601197

GARAGES
Tabley
Tabley Hill Ln, Over Tabley
☎(0565) 3241
Bucklow
Bucklow Hill (BL)
☎(0565) 830327
Pickmere Service Station
Park Ln, Pickmere (Vlo)
☎(056589) 3254
Moss Lane
(J W Egerton) Mobberley
☎(036587) 3196

WHERE TO STAY
★**Rose & Crown**
King St, Knutsford
☎(0565) 52366
☆☆**Swan**
Bucklow Hill (GW)
☎(0565) 830295
GH **Longview**
55 Manchester Rd, Knutsford
☎(0565) 2119

WHERE TO EAT
××**La Belle Époch**
60 King St, Knutsford
☎(0565) 3060
××**David's Place**
10 Princess St, Knutsford
☎(0565) 3356
Sir Frederick's Wine Bar
44 King St, Knutsford
☎(0565) 53209

GARAGES
Tabley
Tabley Hill Ln, Over Tabley
☎(0565) 3241

On its northern side is Tatton Park (NT), a late 18th-century mansion in a 1000-acre park.

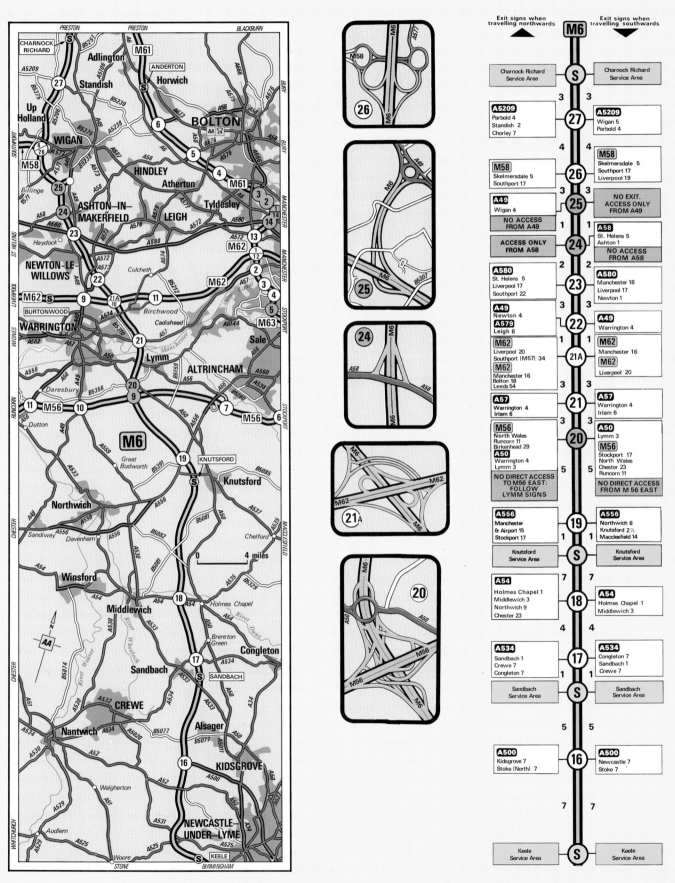

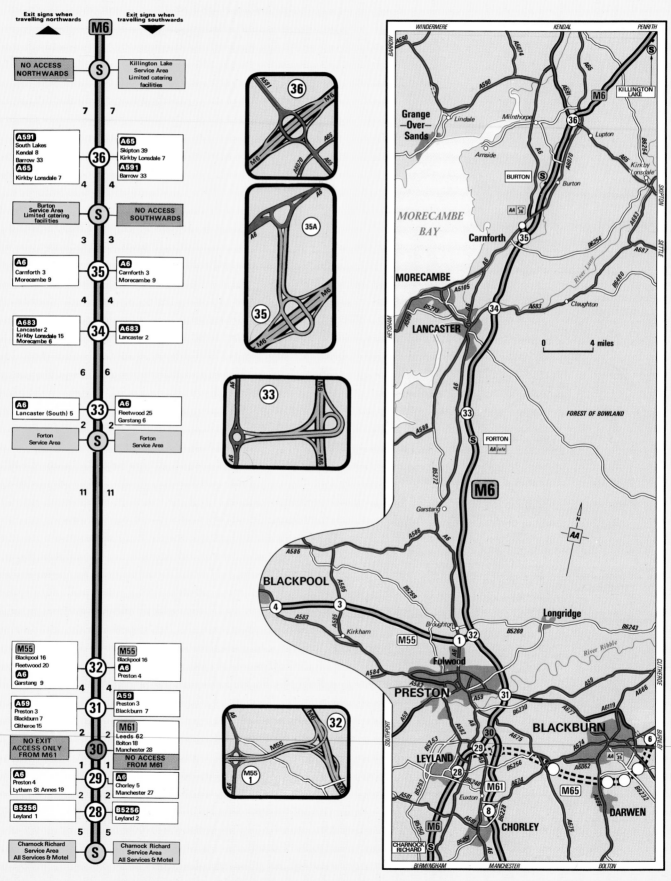

Exit signs when travelling northwards ▲

Exit signs when travelling southwards ▼

M6

NO ACCESS NORTHWARDS — S — Killington Lake Service Area Limited catering facilities

7 | 7

A591 South Lakes Kendal 8 Barrow 33 / A65 Kirkby Lonsdale 7 — 36 — A65 Skipton 39 Kirkby Lonsdale 7 / A591 Barrow 33

4 | 4

Burton Service Area Limited catering facilities — S — NO ACCESS SOUTHWARDS

3 | 3

A6 Carnforth 3 Morecambe 9 — 35 — A6 Carnforth 3 Morecambe 9

4 | 4

A683 Lancaster 2 Kirkby Lonsdale 15 Morecambe 6 — 34 — A683 Lancaster 2

6 | 6

A6 Lancaster (South) 5 — 33 — A6 Fleetwood 25 Garstang 6

2 | 2

Forton Service Area — S — Forton Service Area

11 | 11

M55 Blackpool 16 Fleetwood 20 / A6 Garstang 9 — 32 — M55 Blackpool 16 / A6 Preston 4

4 | 4

A59 Preston 3 Blackburn 7 Clitheroe 15 — 31 — A59 Preston 3 Blackburn 7

2 | 2

NO EXIT ACCESS ONLY FROM M61 — 30 — M61 Leeds 62 Bolton 18 Manchester 28 / NO ACCESS FROM M61

1 | 1

A6 Preston 4 Lytham St Annes 19 — 29 — A6 Chorley 5 Manchester 27

2 | 2

B5256 Leyland 1 — 28 — B5256 Leyland 2

5 | 5

Charnock Richard Service Area All Services & Motel — S — Charnock Richard Service Area All Services & Motel

36

35A

35

33

32

Charnock Richard Services
– see page 68

Junction 28

The town of Leyland is best known for its motor works. At Tarleton farther west are the Leisure Lakes with boating and sailboard facilities amidst heath and woodland.

WHERE TO STAY
★★★**Pines**
Clayton-le-Woods
☎(0772) 38551
☆☆☆**Ladbroke**
(Conferencentre) Leyland Way, Leyland (LB)
☎(07744) 22922
★★★**Shaw Hill Golf and Country Club**
Preston Rd, Whittle-le-Woods
☎(02572) 69221

GARAGES
Swansey
350 Preston Rd, Clayton-le-Woods ☎(02572) 2461
Golden Hill
208 Golden Hill Ln, Leyland (Peu) ☎(07744) 23416
Leyland Service Station
Wigan Rd, Leyland (Sko)
☎(07744) 21546
Thorntree
(M Theobald) Wigan Rd, Leyland ☎(07744) 29400
Chorley Service Station
Harpers Ln, Chorley
☎(02572) 63542
LAC
Portland St, Chorley
☎(02572) 63027

Junction 29

When it is opened the M65 will branch off here for Colne – see page 95 for details. To the east is Houghton Tower, where James I 'knighted' a loin of beef in 1617, thus creating Sirloin.

Junction 30 R

The M61 from Manchester links with the M6 here – see page 87 for details.

Junction 31

Preston, with its docks on the River Ribble, has long been an important manufacturing town, but also has a long and varied history, illustrated by displays in the Harris Museum and Art Gallery. It was also the birthplace of the poet Robert Service, known as the 'Canadian Kipling'.

WHERE TO STAY
☆☆☆**Tickled Trout**
Preston New Rd, Samlesbury
☎(077477) 671
☆☆☆**Trafalgar**
Preston New Rd, Samlesbury
☎(077477) 351
★★★**Broughton Park House**
Garstang Rd, Broughton, Preston ☎(0772) 864087
☆☆☆**Crest**
The Ringway, Preston (CRH)
☎(0772) 59411
GH **Withy Trees**
175 Garstang Rd, Preston
☎(0772) 717693
GH **Fulwood Park**
49 Watling St Rd, Preston
☎(0772) 718067

GH **Beech Grove**
12 Beech Grove, Ashton, Preston ☎(0772) 729969
GH **Tulketh House**
209 Tulketh Rd, Ashton, Preston ☎(0772) 728096
GH **Lauderdale**
29 Fishergate Hill, Preston
☎(0772) 555460

WHERE TO EAT
✕**French Bistro**
Miller Arcade, Church St, Preston ☎(0772) 53882
Alexanders
Winckley St, Preston
☎(0772) 54302
La Bodega
21 Cannon St, Preston
☎(0772) 52159
Danish Kitchen
Barbecue and Coffee Shop, 10 Lune St, Preston
☎(0772) 22086
Samuel Whitbread
Everdale Ln, Samlesbury
☎(077477) 641

GARAGES
Dutton Forshaw Service Centre
Moor Lane, Preston
(DJ RR BL) ☎(0772) 22111
Key Motors
38–40 Garstang Rd, Preston
☎(0772) 53446
Maitland St
(J C Motors) Maitland St, off New Hall Ln, Preston
☎(0772) 794569
International Service Station
Garstang Rd, Broughton (Frd)
☎(0772) 862601

Junction 32

The M55 heads west for Blackpool from this junction – see below.

M55 Preston–Blackpool

From a somewhat sombre area this motorway heads for the fun capital of Britain.

Junction 1

Fulwood and the north of Preston can be reached from this junction. For additional hotels, restaurants and garages see under M6 Junction 31.

WHERE TO STAY
★★★**Bartle Hall**
Lea Ln, Bartle
☎(0772) 690506
★★★**Barton Grange**
Garstang Rd, Barton
☎(0772) 862551

WHERE TO EAT
The Orchard
Whittingham Ln, Broughton
☎(0772) 862208

GARAGES
West End Motors
350 Blackpool Rd, Preston
(Toy) ☎(0772) 719841

Junction 3

The road north leads to Thornton and to Fleetwood for the Isle of Man ferry.

WHERE TO STAY
FH Mrs T Colligan **High Moor Farm** Weeton
☎(039136) 273

WHERE TO EAT
✕✕**Ledra**
Preston New Rd, Freckleton
☎(0772) 632308
The Ship Inn
Freckleton ☎(0772) 32393

GARAGES
Station Road
55–59 Station Rd, Wesham
☎(0772) 682404
Fylde Motor Co
13–15 Preston St, Kirkham
☎(0772) 683561

Junction 4

It can quite safely be said that there is no other place like Blackpool, a resort which has devoted all its abundant energies into leisure and entertainments. It has seven miles of sandy beaches, three piers, a 40-acre amusement park, ballrooms, discos, bingo etc. etc . . . and, of course, its famous tower which itself incorporates dancing, amusements and a circus.

WHERE TO STAY
★★**Headlands**
New South Prom, Blackpool
☎(0253) 41179
★**Kimberley**
New South Prom, Blackpool
☎(0253) 41184
★★**Cliffs**
Queens Prom, Blackpool
☎(0253) 52388
★★**Chequers**
24 Queens Prom, Blackpool
☎(0253) 56431
☆☆☆**Pembroke**
North Prom, Blackpool
☎(0253) 23434
★★**Carlton**
North Prom, Blackpool
☎(0253) 28966
★★**Claremont**
270 North Prom, Blackpool
☎(0253) 293122
★**Revill's**
North Prom, Blackpool
☎(0253) 25768
GH **Motel Mimosa**
24A Lonsdale Rd, Blackpool
☎(0253) 41906
GH **Arandora Star Private Hotel**
559 New South Prom, Blackpool ☎(0253) 41528
GH **Arosa Hotel**
18–20 Empress Dr, Blackpool ☎(0253) 52555
GH **Berwick Private Hotel**
23 King Edward Av, Blackpool ☎(0253) 51496
GH **Burlees Hotel**
40 Knowle Av, Blackpool
☎(0253) 54535
GH **Cliftonville Hotel**
14 Empress Dr, Blackpool
☎(0253) 51052
GH **Denely Private Hotel**
15 King Edward Av, Blackpool ☎(0253) 52757
GH **Edenfield Private Hotel**
42 King Edward Av, Blackpool ☎(0253) 51538
GH **Lynstead Private Hotel**
40 Knowle Av, Blackpool
☎(0253) 51050
GH **North Mount Private Hotel**
22 King Edward Av, Blackpool
☎(0253) 55937
GH **Sunray Private Hotel**
42 Knowle Av, Blackpool
☎(0253) 51937
GH **Surrey House Hotel**
9 Northumberland Av, Blackpool ☎(0253) 51743

M6 Charnock Richard–Killington Lake   51 miles

The north-west holiday resorts of Southport, Blackpool and Morecambe are along the coast here, while inland are the high fells of the Forest of Bowland.

WHERE TO EAT
The Danish Kitchen
Vernon Humpage, Church St, Blackpool ☎(0253) 24291
Corky's
39–41 The Square, Lytham St Annes ☎(0253) 712513
Lidun Cottage Barbecue
5 Church Rd, Lytham
☎(0253) 736936
Tiggis
Express Bldgs 21–23 Wood St, Lytham ☎(0253) 711481

GARAGES
Jeffersons
Alfred St, Blackpool
☎(0253) 26680
Imperial
Dickson Rd, Blackpool
☎(0253) 29031
Ingham & Wicks
Princess St, Blackpool
☎(0253) 24467
Mac's Motors
Clifton Rd, Marton
☎(0253) 67311
Thomas Motors
Oxford Corner, Whitegate Dr, Blackpool ☎(0253) 63333
Williams Bros
Blackpool Airport, Squires Gate, Blackpool
☎(0253) 45421
Woodheads
Squires Gate, Blackpool
☎(0253) 45544

Forton Services
(Rank) ☎(0524) 791775
Restaurant. Fast Food. HGV cafe. Picnic area. Shop. Petrol. Diesel. Breakdowns. Repairs. HGV parking. Long-term/overnight parking for caravans £3 (max 24hrs) Baby-changing. For disabled: toilets (north side only), ramp, lift. Footbridge. Credit cards accepted.

Junction 33

A short distance to the west is Thurnham Hall which dates from the 13th century. Eastwards lie the fells of the Forest of Bowland and the Wyre Valley.

WHERE TO STAY
★**Foxholes**
Bay Horse ☎(0524) 791237
GH **Oakfield**
Lancaster Rd, Forton
☎(0524) 791630

GARAGES
New Holly
Forton ☎(0524) 791424

Junction 34

The county town of Lancaster has some fine old buildings, particularly Shire Hall and Judges Lodgings, which now houses two museums. The famous Hornsea Pottery has a factory here which can be visited.

WHERE TO STAY
☆☆☆☆**Post House**
Caton Rd, Lancaster (THF)
☎(0524) 65999
★★**Royal Kings Arms**
Market St, Lancaster
☎(0524) 32451
★★★**Elms**
Bare, Morecambe
☎(0524) 411501
★★★**Midland**
Marine Rd, Morcambe
☎(0524) 417180
★★★**Strathmore**
Marine Rd East, Morecambe
☎(0524) 411314
★**Rimington**
70–72 Thornton Rd, Morecambe
☎(0524) 415668

WHERE TO EAT
✕✕**Portofino**
23 Castle Hill, Lancaster
☎(0524) 32388
Country Pantry
Co-operative Department Store, Church St, Lancaster
☎(0524) 64355
Old Brussels
53 Market St, Lancaster
☎(0524) 69177
Squirrels
92 Penny St, Lancaster
☎(0524) 62307

GARAGES
S Marshall & Sons
Reynold St, Off St Georges Quay, Lancaster
☎(0524) 67298
O Rix
88 King St, Lancaster (BL LR Rar DJ) ☎(0524) 32233
Auto Service Recovery
40 Braganzo Way, Lune Industrial Est, Lancaster
☎(0524) 33461
Bulk Rd
(RR & D Sowerby) Lancaster (Tal Peu Fia) ☎(0524) 63373
Glanville Lawrence
Penny St, Lancaster (Vau Cit Opl) ☎(0524) 32444

Junction 35

Carnforth is well-known for its Steamtown Railway Museum. The large number of locomotives on display here include the famous 'Flying Scotsman'.

WHERE TO STAY
★★**Royal Station**
Market St, Carnforth
☎(0524) 732033
★**Silverdale**
Shore Rd, Silverdale
☎(0524) 701206

GARAGES
Carnforth Service Station
(D Barnfield) Lancaster Rd, Carnforth (Yam)
☎(0524) 733422

Wayside
(Charlie Oates) Carnforth (Maz) ☎(0524) 2460
Warton Hall
Main St, Warton (Vau Opl)
☎(0524) 732107
Townend
Sands Ln, Warton
☎(0524) 733837

Burton Services
(Granada) Northbound only ☎(0524) 781234
Restaurant. Fast food. Picnic area. Shop. Petrol. Diesel. Breakdowns and repairs on call. HGV parking. Long-term/overnight parking for caravans £3.50. Baby-changing. For disabled: toilets, ramp. Credit cards – shop, garage, restaurant.

Junction 36

This is the junction for traffic to the South Lakeland – Windermere, Ambleside and the lovely Coniston Water.

WHERE TO STAY
★★★**Crooklands**
Crooklands (BW)
☎(04487) 432
★★★**Blue Bell at Heversham**
Prince's Way, Heversham
☎(04482) 3159
★★ ⚑ **Heaves**
(off A6, ½m from junc with A591) Levens
☎(0448) 60396

GARAGES
Crooklands Motor Co
Crooklands ☎(04487) 414
Canal
(J Atkinson & Son Ltd)
Crooklands, Milnthorpe
☎(04487) 401
Crooklands Mill
Crooklands ☎(04487) 216
Service
(E & D E Towers) Holme
☎(0524) 781321

Killington Lake Services
(Roadchef) Southbound only ☎(0567) 20739
Restaurant. Fast food. Picnic area. Shop. Tourist information. Vending machines. Playground. Petrol. Diesel. Breakdowns. Repairs. HGV parking. Long-term/overnight parking for caravans £5. For disabled: toilets, ramp. Credit cards – garage.

LOCAL RADIO STATIONS

	Medium Wave		VHF/FM
	Metres	kHz	MHz
BBC Radio Lancashire	351	855	96.4
IBA Red Rose Radio	301	999	97.3

M6 Killington Lake – Carlisle   57 miles

This must be one of the most scenic stretches of motorway in the country, following the Lune Valley as it winds its way between the high fells.

Killington Lake Services – see page 71

Carlisle's Citadel is a prominent feature of the city, often mistaken for the castle. It was built in 1810 to incorporate and reconstruct Henry VIII's citadel of 1541.

Junction 37

Sedbergh is a busy market town beneath the Howgill Fells. There is a National Park Centre in Main Street.

WHERE TO EAT
※✕**Castle Dairy**
26 Wildman St, Kendal
☎(0539) 21170
Cherry Tree Restaurant
24 Finkle St, Kendal
☎(0539) 20547

GARAGES
Sedbergh Motor Co
Station Rd, Sedbergh (BL)
☎(0587) 20678
Lakeland Motor Co
(Lakeland Ford) Mintsfeet Rd
South, Kendal (Frd)
☎(0539) 23534
Dutton-Forshaw
84–92 Highgate, Kendal (DJ
LR Rar) ☎(0539) 28800

Junction 38

The 'B' road north provides a most attractive drive across the moors to Appleby. This delightful town on the River Eden is famous for its Horse Fair.

WHERE TO EAT
✕**Gilded Apple**
Orton ☎(05874) 345

Tebay, West Services
(Westmorland
Motorways)
Northbound only.
☎(05874) 680.
Accommodation.
Restaurant. Fast food.
Picnic area. Shop.
Vending machines.
Petrol. Diesel.
Breakdowns. Repairs.
Long-term/overnight
parking for caravans
£2.30. Baby-changing.
For disabled: toilets,
ramp.

Junction 39

More fells and moorland surround the motorway. the road west leads to Shap where there are remains of a Premonstratensian Abbey.

WHERE TO STAY
GH **Brookfield**
Shap ☎(09316) 397
FH F Hodgson **Green Farm**
Shap ☎(09316) 619
FH S J Thompson **Southfield Farm** Shap ☎(09316) 282

GARAGES
T Simpson & Sons
Shap (BL) ☎(09316) 212

Junction 40

The lovely old market town of Penrith has a ruined castle at its centre, but much finer is Brougham Castle just outside the town to the south east. South is the Lowther Wildlife Park and west, near Dacre, is Dalemain, a house built around a medieval pele tower which now contains the Westmorland and Cumberland Yeomanry Museum. Ullswater is nearby.

WHERE TO STAY
★★**George**
Devonshire St, Penrith
☎(0768) 62696
★★**Strickland**
Corney Sq, Penrith
☎(0768) 62262
★**Abbotsford**
Wordsworth St, Penrith
☎(0768) 63940
★**Glen Cottage**
Corney Sq, Penrith
☎(0768) 62221
★**Station**
Castlegate, Penrith
☎(0768) 62072

WHERE TO EAT
Waverley
Crown Sq, Penrith
☎(0768) 63962

GARAGES
Ullswater Road
Ullswater Rd, Penrith (Cit Peu
Tal VW Aud) ☎(0768) 64545
Regent
(McDavidson) Tynefield
Bridge Ln, Penrith
☎(0768) 62594
County Motors
Southend Rd, Penrith (BL
Dai)) ☎(0768) 63666
County
Old London Rd, Penrith (Frd)
☎(0768) 64571

Junction 41

About 2 miles north west is Hutton-in-the-Forest, a 14th-century pele tower with later additions. The gardens include an ornamental lake and a nature trail.

Southwaite Services
(Granada)
Restaurant. Fast food
(south side). Picnic area.
Shop. Playground.
Petrol. Diesel.
Breakdowns. Repairs.
HGV parking. Long-
term/overnight parking
for caravans £3.50.
Baby-changing. For
disabled: toilets.
Footbridge. Credit cards
– shop, garage,
restaurant.

Junction 42

The road north-west follows the old Roman road into Carlisle. North east is Wetheral and the nearby Corby Castle, an ancient pele tower around which is a 17th- to 18th-century mansion.

The Howgill Fells near Sedbergh reach a height of over 2000ft and offer superb walking country.

★★**Crown**
Eamont Bridge, Penrith
☎(0768) 62566
★★**Clifton Hill**
Clifton, Penrith
☎(0768) 62717
⚑♨❀★★★**Sharrow Bay**
Sharrow Bay, Pooley Bridge
☎(08536) 301
GH **Brandelhow**
1 Portland Pl, Penrith
☎(0768) 64470
GH **Pategill Villas**
Carleton Rd, Penrith
☎(0768) 63153
GH **Woodland House**
Wordsworth St, Penrith
☎(0768) 64177
GH **Limes**
Redhills, Stainton
☎(0768) 63343
FH Mrs A Strong, **Barton Hall
Farm**
Pooley Bridge ☎(08536) 275

The 'Lady of the Lake' transports holidaymakers across lovely Ullswater.

WHERE TO STAY
★★★**Crown**
Wetheral (BW)
☎(0228) 61888
★★**Killoran**
The Green, Wetheral
☎(0228) 60200
⚑♨★★★**Dalston Hall**
Dalston Rd, Dalston (BW)
☎(0228) 710271

Junction 43

Carlisle is Cumbria's principal town and contains many interesting buildings and museums. Hadrian's wall ran to the north, but the best remains are farther east beyond Brampton.

WHERE TO STAY
★★★**Crown and Mitre**
English St, Carlisle
(COMFORT) ☎(0228) 25491
★★★**Cumbrian Thistle**
Court Sq, Carlisle (TS)
☎(0228) 31951
★★★**Swallow Hilltop**
London Rd, Carlisle (SW)
☎(0228) 29255

★★**Central**
Victoria Viaduct, Carlisle
☎(0228) 20256
★★**Pinegrove**
262 London Rd, Carlisle
☎(0228) 24828
★**Vallum House**
Burgh Rd, Carlisle
☎(0228) 21860
★★**Cumbria Park**
32 Scotland Rd, Carlisle
☎(0228) 22887
★★★**Newby Grange**
Crosby-on-Eden
☎(022873) 645
⚑♨★**Crosby Lodge**
Crosby-on-Eden
☎(022873) 618
GH **Angus**
14 Scotland Rd, Carlisle
☎(0228) 23546
GH **East View**
110 Warwick Rd, Carlisle
☎(0228) 22112
GH **Kenilworth**
24 Lazonby Ter, Carlisle
☎(0228) 26179

WHERE TO EAT
Citadel Restaurant
77–79 English St, Carlisle
☎(0228) 21298
Malt Shovel
Rickergate, Carlisle
☎(0228) 34095

GARAGES
Carleton Service Station
(T Wrathall) Carleton, Carlisle
(Col FSO) ☎(0228) 27287
Carlisle Auto Volks
Millrace Rd, Willowholme
Industrial Est, Carlisle
☎(0228) 46934
County Motors
Montgomery Way, Rose Hill
Est, Carlisle (BL DJ LR Rar)
☎(0228) 24387
Grahams
(Motor Eng) London Rd,
Carlisle ☎(0228) 21739
Robert Street Motor Eng
Robert St, Carlisle
☎(0228) 38323
Corby Hill Motors
(M Appleton) Carlisle
☎(0228) 61308

Junction 44

Here the motorway ends and the A74(T) continues the short distance to the Scottish border and Gretna.

WHERE TO STAY
☆☆☆**Crest Kingstown**
(CRH) ☎(0228) 31201

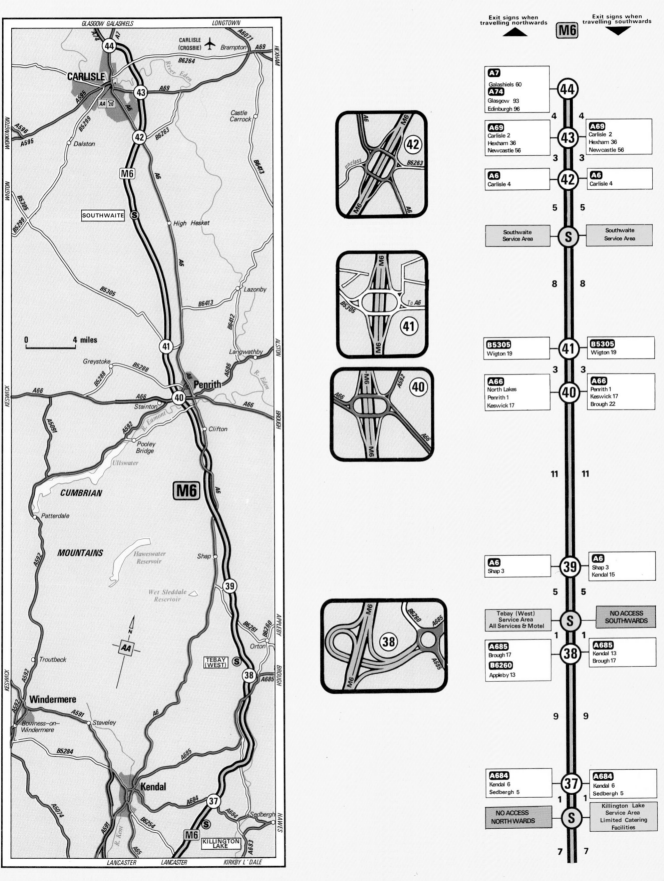

Exit signs when travelling northwards ◀   M6   Exit signs when travelling southwards ▼

Northward	Jct	Southward
A7 Galashiels 60 **A74** Glasgow 93 Edinburgh 96	44	
	4 4	
A69 Carlisle 2 Hexham 36 Newcastle 56	43	**A69** Carlisle 2 Hexham 36 Newcastle 56
	3 3	
A6 Carlisle 4	42	**A6** Carlisle 4
	5 5	
Southwaite Service Area	S	Southwaite Service Area
	8 8	
B5305 Wigton 19	41	**B5305** Wigton 19
	3 3	
A66 North Lakes Penrith 1 Keswick 17	40	**A66** Penrith 1 Keswick 17 Brough 22
	11 11	
A6 Shap 3	39	**A6** Shap 3 Kendal 15
	5 5	
Tebay (West) Service Area All Services & Motel	S	NO ACCESS SOUTHWARDS
	1 1	
A685 Brough 17 **B6260** Appleby 13	38	**A685** Kendal 13 Brough 17
	9 9	
A684 Kendal 6 Sedbergh 5	37	**A684** Kendal 6 Sedbergh 5
	1 1	
NO ACCESS NORTHWARDS	S	Killington Lake Service Area Limited Catering Facilities
	7 7	

73

London Orbital Motorway (West) M25

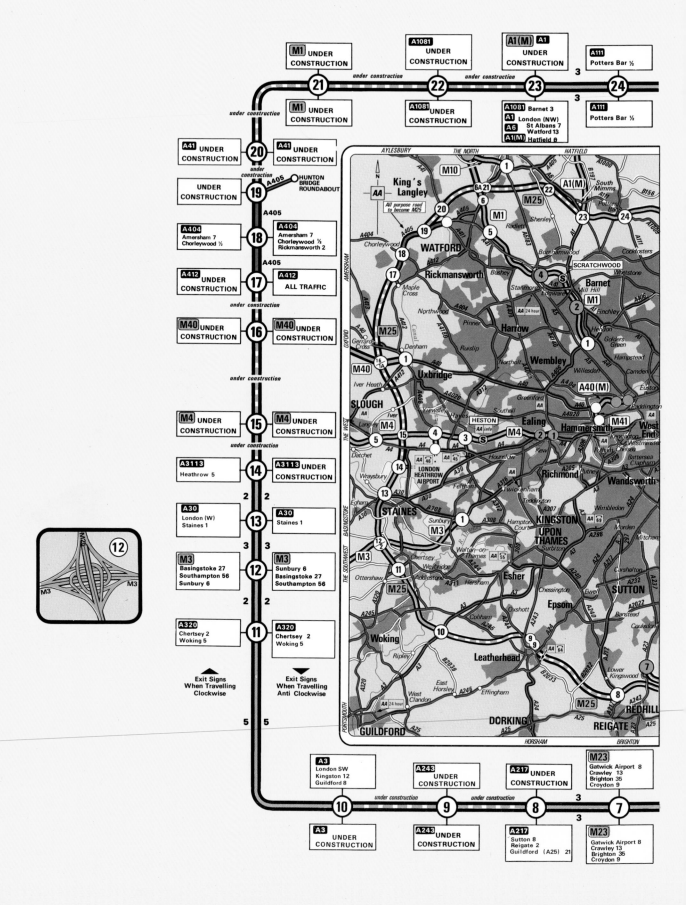

M1 UNDER CONSTRUCTION

A1081 UNDER CONSTRUCTION

A1(M) **A1** UNDER CONSTRUCTION

A111 Potters Bar ½

21

under construction

22

under construction

23

3

24

M1 UNDER CONSTRUCTION

A1081 UNDER CONSTRUCTION

A1081 Barnet 3
A1 London (NW)
A6 St Albans 7
A1(M) Watford 13
Hatfield θ

A111 Potters Bar ½

A41 UNDER CONSTRUCTION

20

A41 UNDER CONSTRUCTION

under construction

UNDER CONSTRUCTION

19

A405 HUNTON BRIDGE ROUNDABOUT

A405

A404 Amersham 7
Chorleywood ½

18

A404 Amersham 7
Chorleywood ½
Rickmansworth 2

A405

A412 UNDER CONSTRUCTION

17

A412 ALL TRAFFIC

under construction

M40 UNDER CONSTRUCTION

16

M40 UNDER CONSTRUCTION

under construction

M4 UNDER CONSTRUCTION

15

M4 UNDER CONSTRUCTION

under construction

A3113 Heathrow 5

14

A3113 UNDER CONSTRUCTION

2   2

A30 London (W)
Staines 1

13

A30 Staines 1

3   3

M3 Basingstoke 27
Southampton 56
Sunbury 6

12

M3 Sunbury 6
Basingstoke 27
Southampton 56

2   2

A320 Chertsey 2
Woking 5

11

A320 Chertsey 2
Woking 5

▲ Exit Signs When Travelling Clockwise

▼ Exit Signs When Travelling Anti Clockwise

5   5

12

A3 London SW
Kingston 12
Guildford 8

A243 UNDER CONSTRUCTION

A217 UNDER CONSTRUCTION

M23 Gatwick Airport 8
Crawley 13
Brighton 35
Croydon 9

10

under construction

9

under construction

8

3

7

A3 UNDER CONSTRUCTION

A243 UNDER CONSTRUCTION

A217 Sutton 8
Reigate 2
Guildford (A25) 21

M23 Gatwick Airport 8
Crawley 13
Brighton 35
Croydon 9

M25 London Orbital Motorway (West)

Although not yet complete, this motorway is already doing much to ease traffic around the outskirts of London. Further details of completion dates etc. can be found on page 24.

The Crest Hotel at South Mimms is situated very conveniently for travellers heading for the Great North Road.

Junction 7

We begin our coverage of this motorway at its junction with the M23, which carries the Gatwick Airport traffic – see page 48 for details.

Junction 8

For the time being the motorway stops here, just north of Reigate. The section between here and junction 10 is under construction and is expected to come into use in the Spring of 1985.

WHERE TO STAY
★★**Reigate Manor**
Reigate Hill, Reigate (BW)
☎(07372) 40125
☆☆**Pickard Motor**
Brighton Rd, Burgh Heath
(BW) ☎(07373) 57222
GH **Cranleigh**
41 West St, Reigate (Minotel)
☎(07372) 40600
GH **Priors Mead**
Blanford Rd, Reigate
☎(07372) 48776
GH **Ashleigh House**
39 Redstone Hill, Redhill
☎(0737) 64763

WHERE TO EAT
❀××**Ebenezer Cottage**
36 Walton St, Walton on the
Hill (073781) 3166
Home Maid
10 Church St, Reigate
☎(07372) 48806
Cobbles Wine Bar
Brewery Yd, Bell St, Reigate
☎(07372) 44833
New Hong Kong
27 Bell St, Reigate
☎(07372) 43374
Crocks
33 High St, Redhill
☎(0737) 61177

GARAGES
Taylor Autos
Reading Arch Rd, Redhill
☎(0757) 62884
South Park Motors
27a Allingham Rd, South
Park, Reigate (Hon)
☎(07372) 43805
Tattenham Corner Motor Company
Ashurst Rd, Tadworth
☎(073781) 2900

Junction 10

Just down the A3 are the famous Wisley Gardens, cultivated by the Royal Horticultural Society. Nearby Woking is a thriving commuter town with the beautiful Shah Jehan Mosque, built in 1889.

WHERE TO STAY
☆☆**Ladbroke Seven Hills**
(and Conferencecentre) Seven
Hills Rd, Cobham (LB)
☎(09326) 4471

WHERE TO EAT
×××**Clock House**
Ripley ☎(0483) 224777

GARAGES
R **Weller**
The Tilt, Cobham (BL)
☎(09326) 4244
Methold Eng
Portsmouth Rd, Ripley
☎(0483) 225373
William Carey Motor Eng
1-2-3 Parvis Rd, West Byfleet
☎(09323) 41073

Junction 11

Chertsey is a pleasant Surrey town retaining some fine Georgian buildings one of which contains the local museum. Nearby is Thorpe Park, a 500-acre theme park.

WHERE TO STAY
★★★**Ship Thistle**
Monument Green, Weybridge
(TS) ☎(0932) 48364
★★★**Oatlands Park**
146 Oaklands Dr, Weybridge
☎(0932) 47242
GH **Warbeck House**
46 Queen's Rd, Weybridge
☎(0932) 48764

WHERE TO EAT
××**Casa Romana**
2 Temple Hall, Monument Hill,
Weybridge ☎(0932) 43470

GARAGES
G **Blair Addlestone**
Chertsey Rd, Addlestone (Tal
Peu) ☎(0932) 42242
Woodham Motors
Woodham Ln, New Haw,
Weybridge (BL)
☎(09323) 42870
Trident
Guildford Rd, Ottershaw (BL
LR) ☎(093287) 2561

Junction 12

The motorway links with the M3 here for Basingstoke, Winchester and Southampton – see pages 46 – 49 for details.

Junction 13

To the west stretches Windsor Great Park with its lovely Savill Garden and Valley Garden. Runnymede is now known for its memorials: to Magna Carta, signed here in 1215; to 20,000 Commonwealth airmen who died in the Second World War; and to President John F Kennedy.

WHERE TO STAY
☆☆☆**Runnymede**
Windsor Rd, Egham
☎(0784) 36171
★★**Pack Horse**
Thames St, Staines (AHT)
☎(0784) 54221

WHERE TO EAT
×××**Bailiwick**
Englefield Green
☎(0784) 32223
××**Terraza**
45 Church St, Ashford
☎(07842) 44887
Maggie's Wine Bar
2 St Judes Rd, Englefield
Green, Egham
☎(0784) 37397
PS **Virginia Water**
2m NE of Sunningdale
OS175 SU9768
PS **Windsor Great Park**
2m S of Windsor
OS175 SU9674

GARAGES
Henlys
Unit 23B, Central Trading Est,
Staines (BL) ☎(0784) 55281
Ayebridges
(MGM Motor Co) Thorpe Lea
Rd, Egham ☎(0784) 56100
Motest
Egham Hill, Egham
☎(0753) 42666
Friary Motors
Straight Rd, Old Windsor (Frd
BL) ☎(07535) 61402
Ashmans
284 – 286 Kingston Rd,
Ashford (Vau)
☎(0784) 52763

Junction 14

Heathrow Airport is a short distance to the east.

WHERE TO STAY
☆☆☆**Sheraton – Heathrow**
Colnbrook Bypass, West
Drayton ☎01 – 759 2424
☆☆☆**Holiday Inn**
Ditton Rd, Langley
(Commonwealth)
☎(0753) 44244

☆☆☆**Excelsior**
Bath Rd, West Drayton (THF)
☎01 – 759 6611
☆☆☆**Heathrow Penta**
Bath Rd, Hounslow
☎01 – 897 6363
☆☆☆**Post House**
Sipson Rd, West Drayton
(THF) ☎01 – 759 2323
★★★**Skyway**
Bath Rd, Hayes (THF)
☎01 – 759 6311

WHERE TO EAT
Chez Petit Laurent
Country Life House, Slough
Rd, Datchet ☎(0753) 49314

GARAGES
Tigwall & Williams
11 David Rd, Poyle Trading
Est, Colnbrook
☎(02812) 2176
Golden Cross Service Station
(MOTEST) Old Bath Rd,
Colnbrook ☎(0753) 42666
Rear of Station
(J C Forrest) Bath Rd,
Colnbrook ☎(08954) 45984

Junctions 15 – 17

This section of motorway is under construction at present and is expected to open during the Spring of 1985.

Junction 17

The A405 between here and junction 19 will eventually be uprated to motorway status – likely to be when the previous section is completed. In anticipation of this, we are treating the junctions here as motorway junctions.

WHERE TO STAY
INN **Greyhound**
High St, Chalfont St Peter
☎(0753) 883404

GARAGES
Mann Egerton & Co
Bury Ln, Rickmansworth (BL
DJ) ☎(0923) 773101

Junction 18

Rickmansworth is now chiefly a residential town. Basing House, in the High Street, was once the home of William Penn, founder of Pennsylvania, USA.

WHERE TO STAY
★★★**Bedford Arms Thistle**
Chenies (TS) ☎(09278) 3301

WHERE TO EAT
Chequers Restaurant
21 Church St, Rickmansworth
☎(0923) 72287
Copper Kettle
Cokes Ln, Little Chalfont
☎(02404) 3144

GARAGES
T A J Motors and Coachworks
33 Station Rd,
Rickmansworth
☎(0923) 773384

Junction 19

This is the junction for Watford. Most of the surviving buildings from old Watford are grouped around the parish church of St Mary.

WHERE TO STAY
☆☆☆**Caledonian**
St Albans Rd, Watford
☎(0923) 29212
GH **White House**
26 – 29 Upton St, Watford
☎(0923) 37316

GARAGES
Hadleigh Ross Autos
101 – 107 Sutton Rd, Watford
(BL) ☎(0923) 26700

Junctions 20 – 23

This section of the motorway is under construction at present and is not due to open until the summer of 1986.

Junction 23

The A1(M) heads northwards from here – see page 105 for details. To the north west, at London Colney, is the Mosquito Aircraft Museum.

WHERE TO STAY
☆☆☆**Crest**
Bignells Corner (junc A1/A6),
South Mimms (CRH)
☎(0707) 43311
☆☆**Elstree Moat House**
Barnet Bypass,
Borehamwood
☎01 – 953 1622

GARAGES
Arlingham Motors
30/38 St Albans Rd, Barnet
☎01 – 449 4036
Ash Motors
Moxon St/High St, Barnet
(Fia) ☎01 – 449 1614

Junction 24

Potters Bar is a modern town which, although close to London, is surrounded by open countryside.

WHERE TO STAY
★★★★**West Lodge Park**
Cockfosters Rd, Hadley
Wood ☎01 – 440 8311
★★★**Ponsbourne**
Newgate Street
☎(0707) 875221

WHERE TO EAT
PS **Trent Park**
½m N of Cockfosters off A111
OS176 TQ2897
PS **Great Wood, Northaw**
1⅓m W of Cuffley on B157
OS166 TL2803

GARAGES
R **Davis**
36 Hatfield Rd, Potters Bar
(BL) ☎(0707) 52371
Cockfosters
347 Cockfosters Rd, Barnet
(Vau Opl) ☎01 – 449 9202
Hadley Green
Great North Rd, Barnet (Tal
Peu RR) ☎01 – 449 5968
Odeon Motors
Great North Rd, Barnet (Sab)
☎01 – 441 2582

A bronze statue of George III on horseback casts a shadow over the Long Walk in Windsor Great Park.

LOCAL RADIO STATIONS			
	Medium Wave		**VHF/FM**
	Metres	*kHz*	*MHz*
BBC Radio London	206	1458	94.9
IBA Capital Radio	194	1548	95.8
IBA LBC	261	1152	97.3

M25 London Orbital Motorway (East)

Only one small section, between Swanley and Sevenoaks in Kent, has yet to be completed on London's eastern bypass. The motorway passes through Epping Forest and the Dartford Tunnel on its way south.

The lovely Epping Forest is a haven for wildlife on the edge of London.

Junction 25

In Enfield, to the south, is Forty Hall, a mansion built in 1629 for the then Lord Mayor of London.

WHERE TO STAY
★★**Holtwhites**
Chase Side, Enfield
☎01–363 0124
★★★**Royal Chace**
The Ridgeway, Enfield
☎01–366 6500

WHERE TO EAT
××**Norfolk**
80 London Rd, Enfield
☎01–363 0979
Blues
41b High St, Waltham Cross
☎(0992) 718633
Divers Wine and Cocktail Bar
29 Silver St, Enfield
☎01–367 2549

GARAGES
Arlington Motor Co
High St, Waltham Cross (Vau Opl) ☎(0992) 760222
Britannia Cross Motors
Britannia Rd, Waltham Cross (BMW)
☎(0992) 712323
ECP Motors
185–189 Turners Hill, Cheshunt (Frd)
☎(0992) 39743
Kilsmore Service Station
(W R Pooley Ltd) 238 Great Cambridge Rd, Cheshunt
☎(0992) 22899
Grinrod & Higgs
Ride Works, Alexandra Rd, Ponders End
☎01–804 4817
Southbury Service Station
101 Southbury Rd, Enfield
☎01–363 8102
Lyne Frank & Wagstaff
London Rd, Enfield (Tal Cit Peu) ☎01–367 3000
Elmsleigh
Redburn Trading Est, Woodall, South St, Ponders End (Frd)
☎01–805 2202
Lavender Hill
The Ridgeway, Enfield (BL)
☎01–363 3456

Junction 26

The motorway passes through the lovely Epping Forest here. To the south, near Chingford, Queen Elizabeth's Hunting Lodge is a picturesque Tudor building housing a museum of the wildlife and history of the forest.

WHERE TO STAY
★★**Roebuck**
North End, Buckhurst Hill (THF) ☎01–505 4636

WHERE TO EAT
Beaton's Wine Bar
319 High St, Epping
☎(0378) 72096

GARAGES
Nick Mason Eng
51 Cartersfield Rd, Waltham Abbey
☎(0992) 713487
Browns
Browns Corner, High Rd, Loughton (Opl Vau)
☎01–508 6262
Wood & Krailing
High Rd, Theydon Bois (Lnc)
☎(037881) 3831
Valley Service Station
1 Valley Hill, Loughton
☎01–508 1787
Station
(J J Tidd) Station Approach, Theydon Bois
☎(037881) 2451

Junction 27

This junction links the M25 with the M11 London–Cambridge motorway here – see page 106 for details.

Junction 28

Brentwood, now a dormitory town for London, has an attractive 428-acre park with lakes and deer at South Weald.

WHERE TO STAY
☆☆☆**Post House**
Brook St, Brentwood (THF)
☎(0277) 210888
★★★★**Brentwood Moat House**
Brentwood (QM)
☎(0277) 225252
GH **Repton**
18 Repton Dr, Gidea Park, Romford
☎(0708) 45253

WHERE TO EAT
Eagle and Child
13 Chelmsford Rd, Shenfield, Brentwood
☎(0277) 210155
PS **Weald Park**
Off A128 at Pilgrims Hatch
OS177 TQ5794

GARAGES
J & R Motor Eng
Howco House, Bryant Av, Romford
☎(04023) 46611

Junction 29

At Upminster to the south west, a 15-century thatched timber building contains the Tithe Barn Agricultural and Folk Museum.

WHERE TO STAY
☆☆☆**Ladbroke**
(& Conferencentre) Southend Arterial Rd, Hornchurch (LB)
☎(04023) 46789

There are many fine old trees in Epping Forest which was a Royal hunting ground for centuries.

WHERE TO EAT
PS **Picnic Site**
1m SE of Brentwood
OS177 TQ6091
PS **Thorndon Country Park**
East Horndon on A128
OS177 TQ6390

GARAGES
HKM Services
Station Rd, West Horndon
☎(0277) 811233
Frost Brothers Motors
68–80 Billet Ln, Hornchurch
(BL) ☎(04024) 46772

Junctions 30 and 31

The motorway approaches the north bank of the Thames here, with the vast container docks at Tilbury to the east. At Grays is the Thurrock Museum of local history. The A282 continues here through the Dartford Tunnel (toll) and on to Dartford.

WHERE TO STAY
☆☆☆**Stifford Moat House**
North Stifford (QM)
☎(0375) 71451

GARAGES
Long Reach Service Station
(D Tongue Autos) London Rd (A13), Purfleet
☎(04026) 4761
Arisdale Motoring Centre
Arisdale Av, South Ockendon
☎(0708) 857335
Easy Cars
Cammavill St, Stifford Clays, Grays ☎(0375) 72737
T Levoi Motors
24–26 Southend Rd, Grays
(Vau Opl) ☎(0375) 76632
Greens
London Rd, Rainham (Vau Opl) ☎(0634) 31242

Junction 1

Although the dual carriageway here is not strictly speaking a motorway, it takes the form of a short link between the existing motorway sections.

GARAGES
Cascade Motor Services
Watling St, Dartford (Tal Peu)
☎(0322) 22166
KT
The Brent, Dartford (Frd)
☎(0322) 22171
RDR Automotive
Ames Rd, Swanscombe
☎(0322) 842510

Junction 2

Just south of this junction, at Sutton-at-Hone, is St John's Jerusalem Garden, bordered by the River Darent and containing the walls and former chapel of the original house to stand on the site.

WHERE TO STAY
☆☆☆**Crest**
Southwold Rd, Bexley (CRH)
☎(0322) 526900

Junction 3

Here at Swanley the motorway links with the M20 – see pages 42–45.

WHERE TO EAT
PS **North Cray Transit Picnic Site**
On A223 1½m S of Bexley
OS177 TQ4871

GARAGES
Dawes
Station Rd, Swanley
☎(0322) 62211
Birchwood Motor Works
Brickwood Corner, Swanley
☎(0322) 63448

Junction 5

The M26 is a short motorway linking this one with the M20 – see pages 42–45 for details. A number of historic houses surround this junction – Knole, famous mansion of the Sackvilles near Sevenoaks; Chartwell, former home of Sir Winston Churchill; Quebec House and Squerries Court, at Westerham, both associated with General Wolfe; Downe House, home of Charles Darwin for 40 years. Emmetts Garden at Brasted is also open.

WHERE TO STAY
★★**Sevenoaks Park**
Seal Hollow Rd, Sevenoaks
☎(0732) 454245
GH **Moorings**
97 Hitchin Hatch Ln, Sevenoaks (Minotel)
☎(0732) 452589

WHERE TO EAT
×××**Ristorante Montmorency**
Quebec Sq, Westerham
☎(0959) 62139

Henry Wilkinson
26 Market St, Westerham
☎(0959) 64245
PS **Dryhill Quarry Transit Picnic Site**
½m S of A25, SE of Sundridge
OS188 TQ4955

GARAGES
London Road Service Station
Badgers Mount
☎(09597) 558
Retreat
(C R Wood & Son) Polhill, Halstead (BL)
☎(0959) 34247
Waite Bros
Sevenoaks Rd, Otford (BL)
☎(09592) 2236
Morants Motors
134 London Rd, Dunton Green (Frd)
☎(073273) 382
Star Motors
128 Seal Rd, Sevenoaks (Vau Opl) ☎(0732) 451337
Stormont Engineering Co
The Vine, Sevenoaks (Frd)
☎(0732) 459911
White Hart
Tonbridge Rd, Sevenoaks
(Cit) ☎(0732) 453328

Junction 6

To the east is Detillens, an interesting 15-century Wealden Hall House and Biggin Hill Airport is to the north east. The motorway has come a full circle and continues to join the section detailed on the previous page.

WHERE TO STAY
★★**Hoskins**
Station Rd West, Oxted (EXEC)
☎(08833) 2338
INN **Whyte Harte**
Bletchingley
☎(0883) 843231

WHERE TO EAT
×**Golden Bengal**
51 Station Rd East, Oxted
☎(08833) 7373
×**Wine and Tandoori**
111–113 Station Rd East, Oxted ☎(08833) 7459
××**La Bonne Auberge**
Tilburstow Hill, South Godstone
☎(0342) 892318
××**Old Lodge**
High St, Limpsfield
☎(08833) 2996
Old Bell
High St, Oxted
☎(08833) 2181
PS **Coulsdon Common**
1½m NW of Caterham
OS187 TQ3257
PS **Tilburstow Hill Viewpoint**
1½m S of Godstone
OS187 TQ3450

GARAGES
Denyer Motors
204 Godstone Rd, Caterham
☎(0883) 42240
Oliver Taylor Recovery
Coach Station, Banstead Rd, Caterham
☎(0883) 42341

LOCAL RADIO STATIONS	Medium Wave		VHF/FM
	Metres	kHz	MHz
BBC Radio London	206	1458	94.9
IBA Capital Radio	194	1548	95.8
IBA LBC	261	1152	97.3

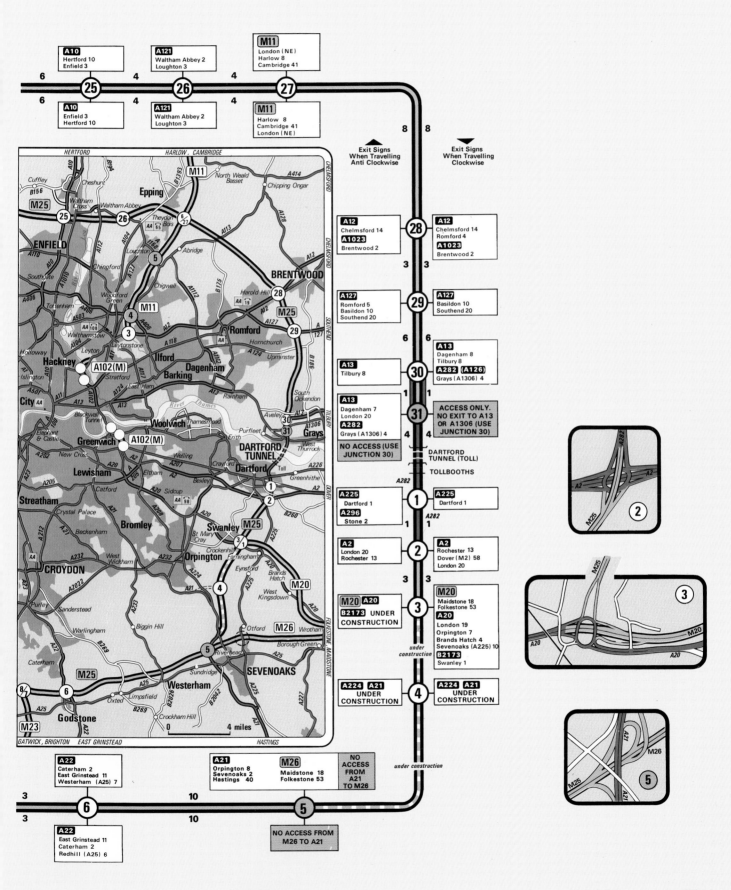

Cadnam - Portsmouth M27

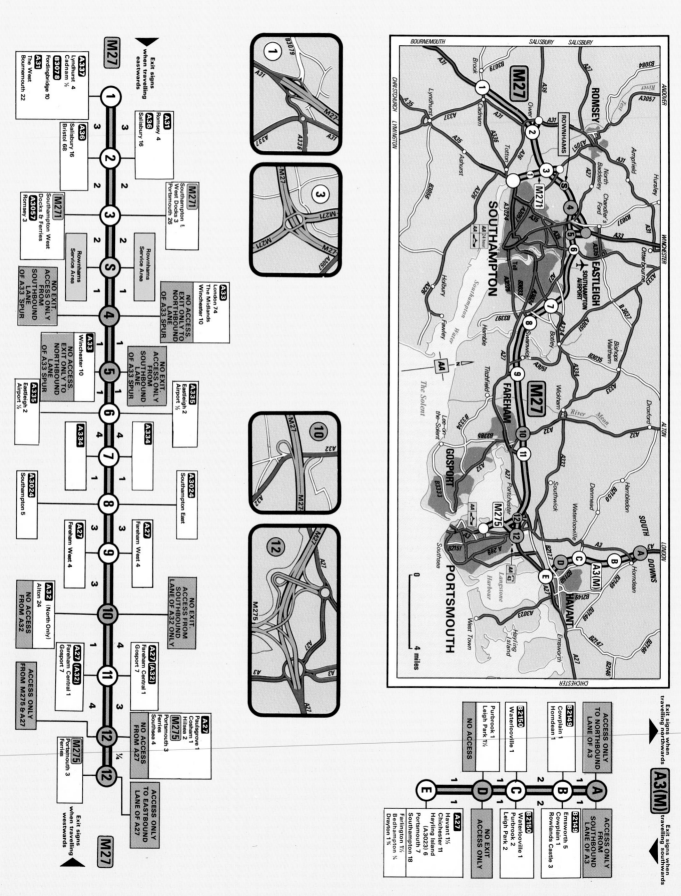

M27

▼ Exit signs when travelling eastwards

1
A337 Lyndhurst 4
Cadnam ½
B3078 Fordingbridge 10
A31 The West
Bournemouth 22

2
A31 Romsey 4
A36 Salisbury 16
A36 Salisbury 16
Bristol 68

3
M271 Southampton W West Docks 3
Portsmouth 26
M271 Southampton West Docks & Ferries
A3057 Romsey 3

S Rownhams Service Area
Rownhams Service Area

4
A33 London 74
The Midlands
Winchester 10
NO ACCESS, EXIT ONLY TO NORTHBOUND LANE OF A33 SPUR
NO EXIT ACCESS ONLY FROM SOUTHBOUND LANE OF A33 SPUR

5
A33 Winchester 10
NO ACCESS, EXIT ONLY TO NORTHBOUND LANE OF A33 SPUR
NO EXIT ACCESS ONLY FROM SOUTHBOUND LANE OF A33 SPUR

6
A335 Eastleigh 2
Airport ½
A335 Eastleigh 2
Airport ½

7
A334
A334

8
A3024 Southampton 5
A3024 Southampton East

9
A27 Fareham West 4
A27 Fareham West 4

10
A32 (North Only) Alton 24
NO ACCESS FROM A32
NO EXIT ACCESS ONLY FROM SOUTHBOUND LANE OF A32 ONLY

11
A27 (A32) Fareham Central 1
Gosport 7
A27 (A32) Fareham Central 1
Gosport 7
ACCESS ONLY FROM M275 & A27
NO ACCESS FROM A27

12
M275 Portsmouth 3
Ferries
A27 Paulsgrove 1
Cosham 1
Hilsea 1
M275 Portsmouth 3
Ferries
Southsea 4
ACCESS ONLY TO EASTBOUND LANE OF A27

12
M275 Portsmouth 3
Ferries

▲ Exit signs when travelling westwards

M27

A3(M)

▶ Exit signs when travelling northwards

ACCESS ONLY TO NORTHBOUND LANE OF A3

B3149 Cowplain 1
Horndean 1

B2150 Waterlooville 1

NO ACCESS

Purbrook 1
Leigh Park 1½

E **D** **C** **B** **A**

B2149 Emsworth 5
Cowplain 1
Rowlands Castle 3

B2150 Waterlooville 1
Purbrook 2
Leigh Park 2

NO EXIT ACCESS ONLY

A27 Havant 1½
Chichester 11
Hayling Island
(A3023) 6
Portsmouth 7
Southampton 18
Farlington 1½
Bedhampton ½
Drayton 1½

ACCESS ONLY FROM SOUTHBOUND LANE OF A3

▶ Exit signs when travelling southwards

78

Junction 1

Within easy reach are Furzey Gardens, the Rufus Stone, Breamore and, of course, Beaulieu.

WHERE TO STAY
⬛★★**Woodlands Lodge**
Bartley Rd, Woodlands
☎(042129) 2257
★★**Bell Inn**
Brook
☎(0703) 812214
★★**Busketts Lawn**
174 Woodlands Rd, Woodlands
☎(042129) 2272
★★**Evergreens**
Romsey Rd, Lyndhurst
☎(042128) 2175
★★**Pikes Hill Forest Lodge**
Pikes Hill, Romsey Rd, Lyndhurst
☎(042128) 3677
★**Forest Point**
Romsey Rd, Lyndhurst
☎(042128) 2420
⬛★★★**Parkhill House**
Lyndhurst
☎(042128) 2944
GH **Barn**
Ashurst
☎(042129) 2531

WHERE TO EAT
××**Le Chanteclerc**
Romsey Rd, Cadnam
☎(042127) 3271
×**Honeysuckle Cottage**
Minstead
☎(0703) 813122
Bow Windows Restaurant
65 High St, Lyndhurst
☎(042128) 2463
PS **Stoney Cross**
2m W of Cadnam off A31
OS195 SU2412
PS **Bolderwood Ornamental Drive**
3m NW of Emery Down
OS195 SU2408

GARAGES
Cadnam
Southampton Rd, Cadnam
(Tal) ☎(0703) 812250
Kibbles
Romsey Rd, Cadnam
☎(0703) 812204
Senior Service Station
Romsey Rd, Cadnam
☎(0703) 813303
Rosary
Brook
☎(0703) 813342

Junction 2

To the north east is Romsey and Broadlands. South, at Totton, the Eling Tide Mill still produces flour.

WHERE TO STAY
★★★**White Horse**
Market Place, Romsey (THF)
☎(0794) 512431
★**Dolphin**
Cornmarket, Romsey
☎(0794) 512171

GARAGES
Netley Marsh
Ringwood Rd, Netley Marsh
☎(0703) 866544

Junction 3

Southampton is Britain's principal commercial port. Among the town's museums is an interesting Maritime Museum.

WHERE TO STAY
GH **Adelaide House**
45 Winchester Rd, Romsey
☎(0794) 512322
GH **Chalet**
Botley Rd, Whitenap, Romsey
☎(0794) 514909
★★★**Southampton Moat House**
119 Highfield Ln, Southampton (QM)
☎(0703) 559555
★★**Albany**
Winn Rd, The Avenue, Southampton
☎(0703) 554553
★★★★**Polygon**
Cumberland Place, Southampton (THF)
☎(0703) 26401
★★★**Southampton Park**
Cumberland Place, Southampton (FD)
☎(0703) 23467
GH **Lodge**
1 Winn Rd, The Avenue, Southampton
☎(0703) 557537
GH **Banister**
11 Brighton Rd, Southampton
☎(0703) 21279
GH **Beacon**
49 Archers Rd, Southampton
☎(0703) 25910

WHERE TO EAT
××**Old Manor House**
21 Palmerston St, Romsey
☎(0794) 517353
××**Olivers**
Ordnance Rd, Southampton
☎(0703) 24789
⊛×**Golden Palace**
Above Bar St, Southampton
☎(0703) 26636
Simon's Wine House
Vernon Walk, Carlton Pl, Southampton
☎(0703) 36372
Pizza Pan
28A Bedford Pl, Southampton
☎(0703) 23103
Mister C's
Park Ln, Off Cumberland Pl, Southampton
☎(0703) 332442

GARAGES
Testwood Motors
331 Salisbury Rd, Totton (Aud VW) ☎(0703) 865727
Lex Tillotson
The Causeway, Redbridge, Southampton (BL)
☎(0703) 860555
Newmans
Redbridge Causeway, Southampton (BL)
☎(0703) 865021
Testwood Motors
Janson Rd, Shirley, Southampton (Aud VW)
☎(0703) 779455
Mitchell Bros
24 Middlebridge St, Romsey (Vau Opl)
☎(0794) 513806
Wrynams
Winchester Rd, Romsey (BL LR) ☎(0794) 512850

Rownhams Services
(Roadchef)
☎(0703) 734480
Restaurant. Fast food. Vending machines. Petrol. Diesel. Breakdowns. Repairs. Long-term/overnight parking for caravans £4. Baby-changing. For disabled: toilets. Tunnel. Credit cards – garage.

Junction 4 R

These junctions link with the A33, the main route northwards

GARAGES
Chandlers Ford Motors
Winchester Ho, School Ln, Chandlers Ford (BL)
☎(04215) 68649
Hendy Lennox
Hendyford Ho, Bournemouth Rd, Chandlers Ford (Frd)
☎(0703) 28331

Junction 5

The local airport at Eastleigh is only a short way from this junction.

WHERE TO STAY
☆☆☆**Crest**
Leigh Rd, Passfield Av, Eastleigh
☎(0703) 619700

GARAGES
F Halfpenny & Son
102 High Rd, Swaythling, Southampton (BL)
☎(0703) 554346
Hampton Park Service Station
306–310 Burgess Rd, Swaythling, Southampton
☎(0703) 554007
Berkeley
21–33 St Denys Rd, Portswood (Dat Peu Tal Frd)
☎(0703) 843036
Woodmill Service Station
(Falcon Motor Services)
Woodmill Ln, Bitterne Park, Southampton
☎(0703) 556483
Alec Bennett
126–152 Portswood Rd, Southampton (Yam Tri Suz Hon Ren)
☎(0703) 554081
Moor Green Service Station
Botley Rd, West End, Southampton
☎(04218) 6247

Junction 7

This junction is on the north-eastern side of Southampton.

GARAGES
Clifton
(Marston Cars) Burnetts Ln, Horton Heath
☎(0703) 692321

Junction 8

The Hamble River, is close to this junction.

WHERE TO EAT
××**Beth's**
The Quay, Hamble
☎(042122) 4314

Junction 9

On a quiet backwater of Portsmouth Harbour is Fareham. Nearby is Titchfield Abbey.

WHERE TO STAY
☆☆☆**Meon Valley**
Sandy Ln, Shedfield
☎(0329) 833455
⊛★★**Old House**
Wickham
☎(0329) 833049
GH **Dormy**
Barnes Ln, Sarisbury Green
☎(04895) 2626

GARAGES
Heath Service Station
Southampton Rd (A27), Titchfield
☎(04895) 3363
Titchfield Motor Works
East St, Titchfield
☎(0329) 43337

M27 Cadnam – Portsmouth   26 miles

The motorway begins within the lovely New Forest and ends at the great naval town of Portsmouth.

Warsash Motors
90 Warsash Rd, Warsash
☎(04895) 5188

Junction 10 R

This junction will take westbound travellers northwards.

GARAGES
Boarhunt
North Boarhunt, Wickham
☎(0329) 833270

Junction 11

To the south, overlooking Portsmouth Harbour, is Portchester Castle.

WHERE TO STAY
★★**Red Lion**
East St, Fareham (WW)
☎(0329) 239611
GH **Maylings Manor**
11A Highlands Rd, Fareham
☎(0329) 286451
GH **Bridgemary Manor**
Brewers Ln, Gosport
☎(0329) 232946

WHERE TO EAT
Gabbies
30–32 West St, Fareham
☎(0329) 284853

GARAGES
Wadham Stringer
(Fareham) East St, Fareham
(BL) ☎(0329) 231511

Wadham Stringer
208–228 West St, Fareham
(BL) ☎(0329) 231511
Youngs
Wickham Rd, Fareham (Frd)
☎(0329) 283221
R & A Ward Bros
Queens Rd, Fareham
☎(0329) 234153
Hillview Motors
Hillview Service Station, 1A Nelson Av, Portchester (Dai)
☎(0705) 374751
Dryad
c/o HMS Dryad, Southwick
☎(0329) 382723
Boarhunt
North Boarhunt, Wickham
☎(0329) 833270

Junction 12 R

The M275 leads into the heart of Portsmouth, Britain's foremost naval port. In the dockyard is the Royal Naval Museum, HMS *Victory*, and the *Mary Rose*.

WHERE TO STAY
★★★**Bear**
East St, Havant (WW)
☎(0705) 486501
GH **Far End**
31 Queens Rd, Waterlooville
☎(07014) 3242

☆☆★**Holiday Inn**
North Harbour, Portsmouth (CHI) ☎(0705) 383151
★★**Keeples Head**
The Hard, Portsmouth (AHT)
☎(0705) 821954
☆☆☆**Crest**
Pembroke Rd, Southsea (CRH)
☎(0705) 827651

WHERE TO EAT
PS **Portsdown**
OS196 SU6606

GARAGES
Ophir Service Station
305–307 London Rd, Hilsea, Portsmouth
☎(0705) 667218
United Services
Vauxhall House, London Rd, Hilsea, Portsmouth (Vau Opl)
☎(0705) 661321
Hendy Lennox
Southampton Rd, Cosham
(Frd) ☎(0705) 370944
Empress
Northern Rd, Cosham (Maz)
☎(0705) 373752
Empress
Havant Rd, Drayton, Portsmouth (Toy)
☎(0705) 374041
Alandown (Motors)
98 West St, Havant
☎(0705) 482369

A3(M) Horndean – Havant

This short section of motorway aids the traffic flow to Portsmouth and Havant from the Petersfield direction.

Junction A R

The A3 becomes motorway here, just south of the beautiful Queen Elizabeth Country Park.

GARAGES
J Watts Autos
South Ln, Clanfield
☎(0705) 595315

Junction B

To the north west is Hambledon, where, it is claimed, the first game of cricket was played in 1774.

WHERE TO EAT
Bat and Ball Inn
Broadhalfpenny Down, Hambledon Rd, Clanfield, Hambledon
☎(070132) 692

GARAGES
Brookside Motor Co
Unit 8, Westfield Industrial Estate, Horndean
☎(0705) 593233
Holmans
249 London Rd, Horndean
☎(0705) 592210
Marshs
(L A Hogett) London Rd, Horndean (PSO)
☎(0705) 593112
H E Hall
The Green, Rowlands Castle (Ren)
☎(070541) 2244

Junctions C and D R

Havant lies to the east, a residential town with some manufacturing industry.

WHERE TO STAY
GH **Far End**
31 Queens Rd, Waterlooville
☎(07014) 3242
FH Mr and Mrs E D Edgell,
Tibbalds Mead Farm White Chimney Row, Westbourne
☎(02434) 4786

GARAGES
Wadham Stringer
Hambledon Rd, Waterlooville
(BL LR)
☎(07014) 2641
Waterlooville Motors
36 Aston Rd, Waterlooville
(Ren) ☎(07014) 51069

Junction E

Langstone Harbour is right in front and nearby is Hayling Island.

WHERE TO STAY
★★★**Bear**
East St, Havant (WW)
☎(0705) 486501
★★**Brookfield**
Havant Rd, Emsworth
☎(02434) 3363
☆☆☆**Post House**
Northney Rd, Hayling Island (THF)
☎(07016) 5011
★★**Newtown House**
Manor Rd, Hayling Island
☎(07016) 66131
GH **Jingles**
77 Horndean Rd, Emsworth
☎(02434) 3755

WHERE TO EAT
PS **Portsdown**
OS196 SU6606

GARAGES
Alandown (Motors)
98 West St, Havant
☎(0705) 482369
Leigh Park
Dunsbury Way, Leigh Park, Havant (Ren)
☎(0705) 473325
Lillywhite Bros
Queen St, Emsworth
☎(02434) 2336

LOCAL RADIO STATIONS	Medium Wave		VHF/FM
	Metres	kHz	MHz
BBC Radio Solent	300	999	96.1
IBA Radio Victory	257	1170	95.0

M40 Denham–Oxford   31 miles

From London's north-western edge, this motorway crosses the lovely Chiltern Hills to reach Oxford, the university city, famous for its dreaming spires.

The wide river Thames flows under the elegant suspension bridge at Marlow, one of the most attractive towns on the river. A cascading weir completes the scene.

Junction 1

The motorway begins close to that great thoroughfare of the past, the Grand Union Canal.

WHERE TO STAY
☆☆☆**Master Brewer Motel**
Western Av (A40) Hillingdon
☎(0895) 51199
★★★**Bull**
Gerrards Cross (DV)
☎(0753) 885995
GH **Bridgettine Convent**
Fulmer Common Rd, Iver Heath
☎(02816) 2645
GH **Woodlands**
84 Long Ln, Ickenham
☎(08956) 34830
GH **Seventeenth-century Barn**
West End Rd, Ruislip
☎(08956) 36057

WHERE TO EAT
×××**Giovanni's**
Denham Lodge, Oxford Rd, Uxbridge
☎(0895) 31568
PS **Bayhurst Wood**
2m N of Uxbridge
OS176 TQ0688
PS **Mad Bess Wood**
1½m N of Ruislip
OS176 TQ0789

GARAGES
Orbital Service Station
North Orbital Rd, Denham
☎(0895) 832180
Courtwood Car Services
Penfield Est, Lancaster Rd, Uxbridge
☎(0895) 36567
Michael Reeves Motors
33 Belmont Rd, Uxbridge (Maz)
☎(0895) 36565
Mamos
325 Long Ln, Hillingdon (BL)
☎(0895) 53681
Motest
148 West Drayton Rd, Hillingdon
☎01-573 0204

Junction 2

Beaconsfield is a pleasant town, associated with

Disraeli, which has what is reputed to be the oldest model village in the world. At Chalfont St Giles is the cottage where Milton lived and finished his 'Paradise Lost'.

WHERE TO STAY
☆☆**Crest**
Aylesbury End, Beaconsfield (CRH)
☎(04946) 71211
☆☆☆☆**Bellhouse**
(2m E A40) Beaconsfield (DV)
☎(0753) 887211
★★★**Bull**
Gerrards Cross (DV)
☎(0753) 885995
INN **Greyhound**
High St, Chalfont St Peter
☎(0753) 883404

WHERE TO EAT
×**Jasmine**
15a Penn Rd, Beaconsfield
☎(04946) 5335
×**China Diner**
7 The Highway, Station Rd, Beaconsfield (04946) 3345
PS **Chilterns Picnic Place (Hodgemoor Wood)**
⅓m E of A355 OS175 SU9594

GARAGES
Hughes
55 Station Rd, Beaconsfield (Frd MB Toy)
☎(04946) 2141
Devonshire Service Station
Chalfont Rd, Seer Green, Beaconsfield (BL)
☎(04946) 6363

Junction 3 R

The road south crosses the Thames to Maidenhead. Standing above the river is Cliveden, former home of the Astors.

WHERE TO STAY
INN **Royal Exchange**
Cookham
☎(06285) 20085

WHERE TO EAT
×××**Bel & the Dragon**
High St, Cookham
☎(06285) 21263

×**Le Radier**
19–21 Station Hill Pde, Cookham
☎(06285) 25775
The Two Roses
High St, Cookham
☎(06285) 20875

GARAGES
Slades
Penn (BL)
☎(049481) 2115
Barnsides Motors
High St, Cookham
☎(06285) 22029

Junction 4

Famous for its furniture-making industry, High Wycombe is a thriving town. Nearby is West Wycombe Park (NT), rebuilt for Sir Francis Dashwood in 1765, and West Wycombe Caves where he held meetings of his infamous Hell Fire Club. There is also a Motor Museum in West Wycombe.

WHERE TO STAY
☆☆☆**Crest**
Crest Rd, Handy Cross, High Wycombe (CRH)
☎(0494) 442100
★★**Falcon**
High Wycombe
☎(0494) 22173
★★★★**Compleat Angler**
Marlow Bridge, Marlow (THF)
☎(06284) 4444
GH **Amersham Hill**
52 Amersham Hill, High Wycombe
☎(0494) 20635
GH **Drake Court**
London Rd, High Wycombe
☎(0494) 23639
GH **Clifton Lodge**
210–212 West Wycombe Rd, High Wycombe
☎(0494) 40095
GH **Glade Nook**
75 Glade Rd, Marlow
☎(06284) 4677

WHERE TO EAT
××**Cavaliers (Restaurant Français)**
24 West St, Marlow
☎(06284) 2544
Burgers
The Causeway, Marlow
☎(06284) 3389
PS **Marlow–Bisham bypass**
S of Marlow on A40
OS175 SU8584
PS **Marlow, Winter Hill**
On Marlow–Cookham road
OS175 SU8786

GARAGES
Marlow Service Station
41 Marlow Rd, High Wycombe
☎(0494) 22888
Davenport Vernon & Co
London Rd, High Wycombe (Dat Vau Opl)
☎(0494) 30021
Rye Mill
70–76 London Rd, High Wycombe (Aud VW Kaw)
☎(0494) 30005
Platts
(R J E Platt Ltd) West St, Oxford Rd, Marlow (BL MG)
☎(06284) 6333
Turnpike
(T S S Booker) 246 New Rd, Booker
☎(0494) 23471
G I Thomas
Unit 3, Abercrombie Industrial Est, High Wycombe
☎(0494) 37414

Junction 5

This junction is high up in the lovely Chilterns. To the south is Stonor House and Park originally built c1190 and home of the Stonor family for 800 years.

WHERE TO EAT
PS **Cowleaze Wood**
Off A40, 2m W of Stokenchurch
OS165 SU7295

GARAGES
Five Alls
Oxford Rd, Studley Green, Stokenchurch
☎(024026) 3200
Tower
Oxford Rd, Stokenchurch
☎(024026) 3355

Junction 6

The narrow lanes around this junction offer some of the most scenic drives in the Chilterns.

GARAGES
Postcombe Service Station
(A40) London Rd, Postcombe
☎(084428) 222

Junction 7 R

The motorway ends here and dual carriageway continues towards Oxford.

WHERE TO STAY
★★**Belfrey**
Brimpton Grange, Milton Common
☎(08446) 381
FH S Fonge **Manor Farm**
Waterperry
☎(08447) 263

GARAGES
Three Pigeons
(H & B Motors) Milton Common
☎(08446) 648

OXFORD
A modern shopping centre and the sprawling motor works have not robbed Oxford of its air of tranquillity and antiquity. Its atmosphere is due to the fact that at its heart lie the ancient buildings which are the home of England's oldest university. Here the bustle is left behind in the quiet cloisters and quadrangles of the colleges that have remained unchanged for hundreds of years. Here too is England's oldest museum, the Ashmolean, which contains one of the finest art collections in the country and many other treasures. Other museums include History of Science, Modern Art, Antique Dolls' Houses and the Museum of Oxford. The University Private Botanic Garden in the High Street and the University Arboretum, 5 miles south-east at Nuneham Courtenay, are open to the public. Details of access to the colleges can be obtained from the Official Information Bureau, Carfax Tower.

WHERE TO STAY
★★★★**Randolph**
Beaumont St (THF)
☎(0865) 247481
★★★**Ladbroke Linton Lodge**
Linton Rd, Park Town (L)
☎(0865) 53461
☆☆☆**Oxford Moat House**
Godstow Rd, Wolvercote Rbt (Q) ☎(0865) 59933
☆☆☆**TraveLodge**
Peartree Rbt (THF)
☎(0865) 54301
★★**Cotswold Lodge**
66A Banbury Rd
☎(0865) 512121
★★**Eastgate**
The High, Merton St (A)
☎(0865) 248244
★★**Isis**
47–53 Iffley Rd
☎(0865) 248894
★★**Royal Oxford**
Park End St (E)
☎(0865) 248432
★**River**
17 Botley Rd
☎(0865) 243475
GH **Ascot**
283 Iffley Rd
☎(0865) 240259
GH **Brown's**
281 Iffley Rd
☎(0865) 246822
GH **Combermere**
11 Polstead Rd
☎(0865) 56971
GH **Conifer**
116 The Slade, Headington
☎(0865) 63055

GH **Earlmont**
322–324 Cowley Rd
☎(0865) 240236
GH **Falcon**
88–90 Abingdon Rd
☎(0865) 7229955
GH **Galaxie Private Hotel**
180 Banbury Rd
☎(0865) 55688
GH **Micklewood**
331 Cowley Rd
☎(0865) 247328
GH **Pine Castle**
290 Iffley Rd
☎(0865) 241497
GH **St Giles Hotel**
56 St Giles
☎(0865) 54620
GH **Tilbury Lodge**
5 Tilbury Ln, Botley
☎(0865) 862138
GH **Victoria Hotel**
180 Abingdon Rd
☎(0865) 724536
GH **Westgate Hotel**
1 Botley Rd
☎(0865) 726721
GH **Westwood Country Hotel**
Hinksey Hill Top (M)
☎(0865) 735408
GH **Willow Reaches Private Hotel**
1 Wytham St
☎(0865) 721545

WHERE TO EAT
❀×××**Restaurant Elizabeth**
84 St Aldates
☎(0865) 242230
××**Paddyfield**
39–40 Hyther Bridge St
☎(0865) 248835
×**Clements**
37 St Clements
☎(0865) 241431
×**Lotus House**
197 Banbury Rd, Parktown, Summertown
☎(0865) 54239
×**Michel's Bistro**
146 London Rd, Headington
☎(0865) 62587
×**Opium Den**
79 George St
☎(0865) 248680
❀×**Wren's**
29 Castle St
☎(0865) 242944
Burlington Bertie's Restaurant and Coffee House
9a High St
☎(0865) 723342
Maxwell's
36 Queen St
☎(0865) 242192
The Nosebag
6–8 St Michael's St
☎(0865) 721033

GARAGES
Groom & Hornsby
496 Cowley Rd
☎(0865) 778282
Hartwells
Watlington Rd, Cowley
☎(0865) 777744
J T Motors
Worcester Pl
☎(0865) 59078
North Oxford
280 Banbury Rd
☎(0865) 511461
St Clements
Dawson St
☎(0865) 244554

LOCAL RADIO STATIONS

	Medium Wave		VHF/FM
	Metres	kHz	MHz
BBC Radio London	206	1458	94.9
BBC Radio Oxford	202	1485	95.2
IBA Capital Radio	194	1548	95.8
IBA LBC	261	1152	97.3

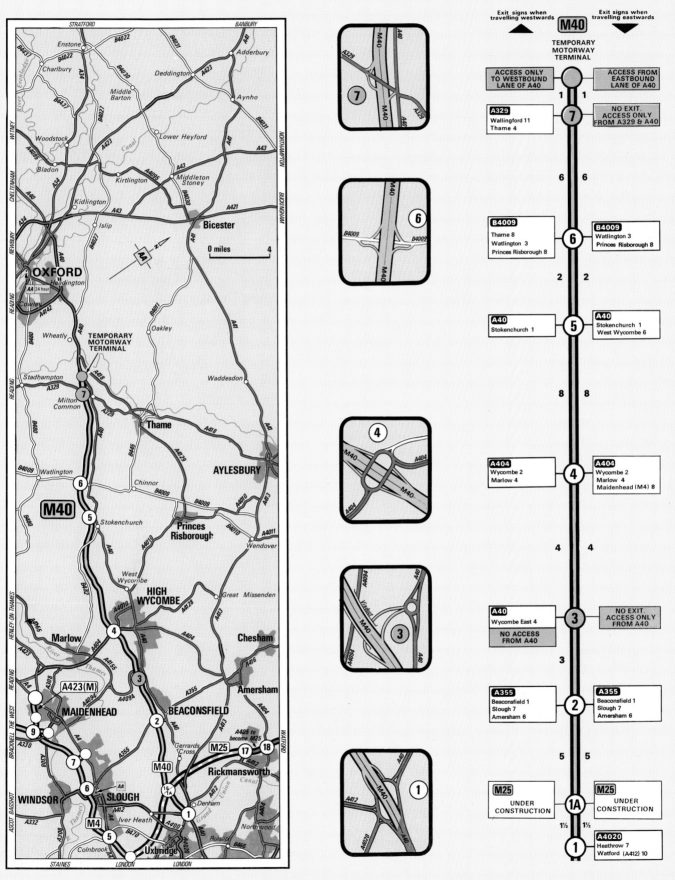

Exit signs when travelling westwards ▲

Exit signs when travelling eastwards ▼

M40

TEMPORARY MOTORWAY TERMINAL

| ACCESS ONLY TO WESTBOUND LANE OF A40 | | ACCESS FROM EASTBOUND LANE OF A40 |

1 1

A329
Wallingford 11
Thame 4

(7)

NO EXIT. ACCESS ONLY FROM A329 & A40

6 6

B4009
Thame 8
Watlington 3
Princes Risborough 8

(6)

B4009
Watlington 3
Princes Risborough 8

2 2

A40
Stokenchurch 1

(5)

A40
Stokenchurch 1
West Wycombe 6

8 8

A404
Wycombe 2
Marlow 4

(4)

A404
Wycombe 2
Marlow 4
Maidenhead (M4) 8

4 4

A40
Wycombe East 4

NO ACCESS FROM A40

(3)

NO EXIT. ACCESS ONLY FROM A40

3 3

A355
Beaconsfield 1
Slough 7
Amersham 6

(2)

A355
Beaconsfield 1
Slough 7
Amersham 6

5 5

M25
UNDER CONSTRUCTION

(1A)

M25
UNDER CONSTRUCTION

1½ 1½

A4020
Heathrow 7
Watford (A412) 10

(1)

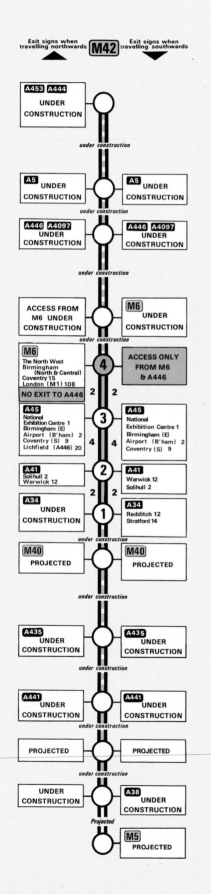

Exit signs when travelling northwards ▲ M42 Exit signs when travelling southwards ▼

A453 A444
UNDER CONSTRUCTION

under construction

A5 UNDER CONSTRUCTION — A5 UNDER CONSTRUCTION

under construction

A446 A4097 UNDER CONSTRUCTION — A446 A4097 UNDER CONSTRUCTION

under construction

ACCESS FROM M6 UNDER CONSTRUCTION — M6 UNDER CONSTRUCTION

under construction

M6
The North West
Birmingham (North & Central)
Coventry 15
London (M1) 108
NO EXIT TO A446
— (4) —
ACCESS ONLY FROM M6 & A446

2   2

A45
National Exhibition Centre 1
Birmingham (E)
Airport (B'ham) 2
Coventry (S) 9
Lichfield (A446) 20
— (3) —
A45
National Exhibition Centre 1
Birmingham (E)
Airport (B'ham) 2
Coventry (S) 9

4   4

A41
Solihull 2
Warwick 12
— (2) —
A41
Warwick 12
Solihull 2

2   2

A34 UNDER CONSTRUCTION — (1) — A34
Redditch 12
Stratford 14

under construction

M40 PROJECTED — M40 PROJECTED

under construction

A435 UNDER CONSTRUCTION — A435 UNDER CONSTRUCTION

under construction

A441 UNDER CONSTRUCTION — A441 UNDER CONSTRUCTION

under construction

PROJECTED — PROJECTED

under construction

UNDER CONSTRUCTION — A38 UNDER CONSTRUCTION

Projected

M5 PROJECTED

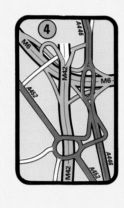

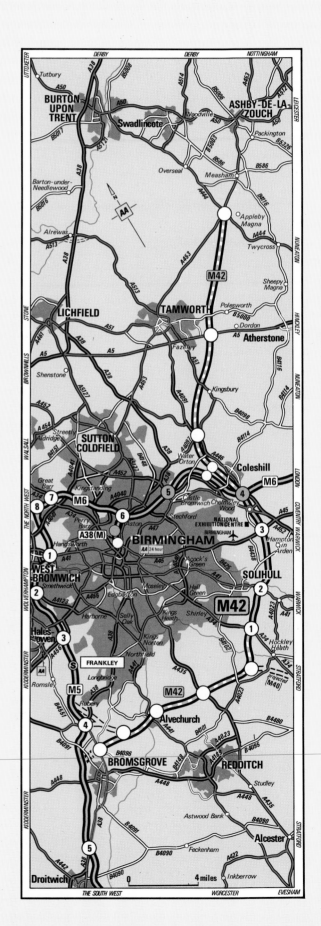

Junction 1

Birmingham is to the north-west and *en route* to the city centre is Sarehole Mill, an 18th-century water mill, restored to working order. South-east of the junction is Packwood House, a timber-framed 16th- to 17th-century house with a famous garden in which yews are clipped to represent the Sermon on the Mount.

WHERE TO STAY
★★*B* **Forest**
Station Approach, Dorridge
☎(05645) 2120

WHERE TO EAT
✕**La Villa Bianca**
1036 Stratford Rd, Monkspath, Shirley
☎021–744 7232

GARAGES
M Grimes
Four Ashes Rd, Dorridge (Dat) ☎(05645) 5118
Cranmore
Drayton St, Shirley, Solihull
☎021–704 1181
Earlswood
The Common, Earlswood (Rel) ☎(05646) 2254
Bristol Street Motors
Stratford Rd, Shirley (Frd)
☎021–744 4456
Willpower
Old Warwick Rd, Lapworth
☎(05643) 2374

Junction 2

The Birmingham Railway Museum is on the road from this junction towards the city, a working museum displaying steam locomotives and historic carriages, wagons etc in their true environment

WHERE TO STAY
★★*B* **Greswolde Arms**
High St, Knowle (AB)
☎(05645) 2711
★★★**George**
High St, Solihull (EH)
☎021–704 1241
★★★**St Johns**
651 Warwick Rd, Solihull (IH)
☎021–705 6777
★★**Flemings**
141 Warwick Rd, Olton, Solihull ☎021–706 0371

WHERE TO EAT
Ye Olde Bakehouse
Warwick Rd, Chadwick End, Knowle ☎(05643) 2928
Bobby Brown's
183 High St, Solihull
☎021–704 9136

GARAGES
Greswolde
(Stirling Motor) Warwick Rd, Knowle ☎(05643) 2341
Damson Lane Service Station
Damson Ln, Solihull
☎021–704 0915

Junction 3

To the west is Birmingham Airport and the huge National Exhibition Centre (see right). Closer to the city centre, at Yardley, is Blakesley Hall, a timber-framed yeoman's house of c1575, now partly furnished in 17th-century style.

WHERE TO STAY
☆☆☆**Arden Motel**
Coventry Rd, Bickenhall Village ☎(06755) 3221
★★★★**Excelsior**
Coventry Rd, Elmdon (THF)
☎021–743 8141
★★**Wheatsheaf**
Coventry Rd, Sheldon (AB)
☎021–743 2021
★★★**Manor**
Meriden (DV) ☎(0676) 22735
GH **Meriden**
Main Rd, Meriden
☎(0676) 22005
GH **Tri-Star**
Coventry Rd, Elmdon
☎021–779 2233

GARAGES
Kenilworth Road
(Kenilworth Motor Racing Services) Hampton-in-Arden
☎(06755) 2292
Ring of Bells
Solihull Rd, Hampton-in-Arden (Vlo)
☎(06755) 2288
Skyways Service Station
Coventry Rd, Birmingham Airport ☎021–742 8943

Junction 4 R

The M42 links with the M6 here and this is as far as it goes for the present. Not until the summer of 1986 will it continue northwards to Appleby Magna.

WHERE TO STAY
★★**Swan**
High St, Coleshill (AB)
☎(0675) 62212
GH **Heath Lodge**
Coleshill Rd, Marston Green
☎021–779 2218
INN **George & Dragon**
154 Coventry Rd, Coleshill
☎(0675) 62249

GARAGES
Chelmsley Wood
Chester Rd, Bacons End, Chelmsley Wood, (Ren)
☎021–770 8373
Murco Service Station
Greenlands Rd, Chelmsley Wood ☎021–770 5459

M42 Bromsgrove–Appleby Magna   8 miles open

At the time of going to press the greater part of this motorway is still under construction.

THE NATIONAL EXHIBITION CENTRE

This new exhibition centre was opened by HM The Queen in 1976 and it could not have been better placed for easy access from all parts of the country. It lies at the very heart of our motorway system, has its own railway station and is next door to Birmingham Airport.

The complex includes nine exhibition halls, with an interior area of 100,985 sq metres, which are centred on the Piazza. Here, services to visitors include shops, banks, medical services and an Information Bureau. The buildings are surrounded by landscaped grounds, including the man-made Pendigo Lake, which cover 310 acres. Around this, parking has been provided for 15,000 cars and 200 coaches. Parking fees depend on the exhibition being staged, but there is a free shuttle bus service to the main entrance. Two hotels (not AA-appointed) are also included, one of which provides conference and banqueting facilities. For AA listed hotels see under M42 Junction 3.

With one or two notable exceptions, the majority of the exhibitions held here are trade shows and are not open to the general public. A list of exhibitions planned is available from the National Exhibition Centre Ltd, Birmingham B40 1NT ☎021–780 4141.

Although Solihull has a great deal of modern building, including a new shopping precinct, many of its older streets retain their character. The High Street has a 15th-century manor house and the George Hotel (below), once a coaching inn, which overlooks a medieval bowling green.

M54 Telford Motorway   23 miles

From the M6 north of Walsall, this motorway crosses rural Staffordshire and Shropshire to Telford and the Wrekin.

Junction 1

Beside the motorway to the west is Moseley Old Hall, built in Elizabethan times and one-time refuge of Charles II after the Battle of Worcester in 1651.

GARAGES
Hilton Service Station
Cannock Rd, Featherstone
☎(0902) 732566
Wednesfield Auto Centre
Patrick Gregory Rd, Wednesfield
☎(0902) 733425

Junction 2

The road south leads to Wolverhampton. To the west is Chillington Hall, an 18th-century house in landscaped grounds by 'Capability' Brown. Nearby is Boscobel House where the grounds contain a descendent of the famous Royal Oak, in which Charles II hid after his defeat at Worcester.

Views can extend over a dozen counties from the summit of The Wrekin near Telford.

M69 Coventry – Leicester   16 miles

This short stretch of motorway links not only the two cities, but also two motorways – the M6 and the M1.

Junction 1

Hinckley has a long history of hosiery manufacturing and it was here in 1640 that one of the first British stocking frames was installed. North of the town is Bosworth Field where Richard III fell in 1485.

WHERE TO STAY
GH **Cecilia's**
13–19 Mount Rd, Hinckley
☎(0455) 637193

WHERE TO EAT
PS **Burbage Common**
1m W of Hinckley OS140 SP4495

Junction 2 R

Another access to Hinckley, but only for southbound traffic. To the north is the Cadeby Light Railway, probably the smallest of Britain's narrow gauge railways.

WHERE TO STAY
★**Fernleigh**
32 Wood St, Earl Shilton
☎(0455) 47011

See also M1 junction 21, page 36, and M6 junction 2, page 64

LOCAL RADIO STATIONS

	Medium Wave		VHF/FM
	Metres	kHz	MHz
BBC Radio Leicester	358	837	95.1
IBA Leicester area	238	1260	97.1

Bushbury Motor Co
70–88 Stafford Rd, Wolverhampton
☎(0902) 28998
Wolverhampton Motor Bodies
Woden Rd, Heath Town, Wolverhampton
☎(0902) 55240
Brewood Service
Coven Rd, Brewood (BL)
☎(0902) 850224
Gorsty Lea
Wolverhampton Rd, Codsall
☎(09074) 2591

Junction 3

RAF Cosford is just to the south, where an Aerospace Museum has an extensive collection of aircraft, engines, rockets and missiles. Weston Park, to the north, is a fine 17th-century mansion in landscaped grounds.

WHERE TO EAT
✕✕✕**Lea Manor**
Holyhead Rd, Albrighton
☎(090722) 3266

Junction 4

Shifnal, to the south-east, is an old town with many interesting buildings. It was mentioned by Charles Dickens in his 'The Old Curiosity Shop'.

WHERE TO STAY
★★★H **Park House**
Park St, Shifnal
☎(0952) 460128
☆☆☆B **Telford Hotel Golf and Country Club**
Great Hay, Sutton Hill
☎(0952) 585642

WHERE TO EAT
Le Bistro Steakhouse
6 School St, Wolverhampton
☎(0902) 24638
Pepito's
5 School St, Wolverhampton
☎(0902) 23403

GARAGES
Croft
(E W & P Barrow) Brewood Rd, Coven (Sub)
☎(0902) 790217
Birches Bridge
Wolverhampton Rd, Codsall, (BL) ☎(09074) 2316

GARAGES
Cheapside
(Patons) Shifnal (BL)
☎(0952) 460412
Oakleys Agricultural Division
Park St, Shifnal (Frd)
☎(0952) 460631
Furrows
Haygate Rd, Wellington, Telford (Frd)
☎(0952) 42433
Parkway Service Station
(E G & B M Read) Madeley Bypass, Telford
☎(0952) 585961

Junction 5

Telford, England's newest New Town, is steadily growing on both sides of the motorway. Named after Thomas Telford, it has strong links with the Industrial Revolution which is said to have been born just south of here around Coalbrookdale. An extensive Industrial Museum complex is centred on the famous Iron Bridge.

GARAGES
Vincent Greenhouse
Holyhead Rd, Ketley, Telford (Vau Opl) ☎(0952) 618081
Hill Top
(J W Stokes) Arleston Hill, Dawley, Telford
☎(0952) 503100
Telford Service Station
Finger Rd, Dawley, Telford
☎(0952) 595567

LOCAL RADIO STATIONS

	Medium Wave		VHF/FM
	Metres	kHz	MHz
BBC Radio WM	206	1458	95.6
IBA Beacon Radio	303	990	97.2

Junction 6

This junction gives access to the western part of Telford. Part of the planning for this new town was the planting of more than a million trees, bushes and shrubs to soften the outline and create a 'forest city'.

WHERE TO STAY
★★**Charlton Arms**
Wellington, Telford (GW)
☎(0952) 51351
★**White House**
Wellington, Donnington, Telford ☎(0952) 604276

Junction 7

The motorway ends here beneath the wooded slopes of the Wrekin, a 1,334 ft hill with panoramic views. At its summit is an Iron-age Hill Fort.

WHERE TO STAY
★★**Falcon**
Holyhead Rd, Wellington, Telford ☎(0952) 55011
★★★B **Buckatree Hall**
Ercall Ln, Wellington, Telford
☎(0952) 51821
INN **Swan**
Watling St, Wellington
☎(0952) 3781

WHERE TO EAT
PS **Ercall Wood**
1m SSW of Wellington OS127 SJ6409

Kirby Muxloe Castle is to the west of Leicester, a moated brick-built manor house which was built by Lord Hastings during the Wars of the Roses and completed shortly before his execution. The remains are now designated an Ancient Monument and are open every day.

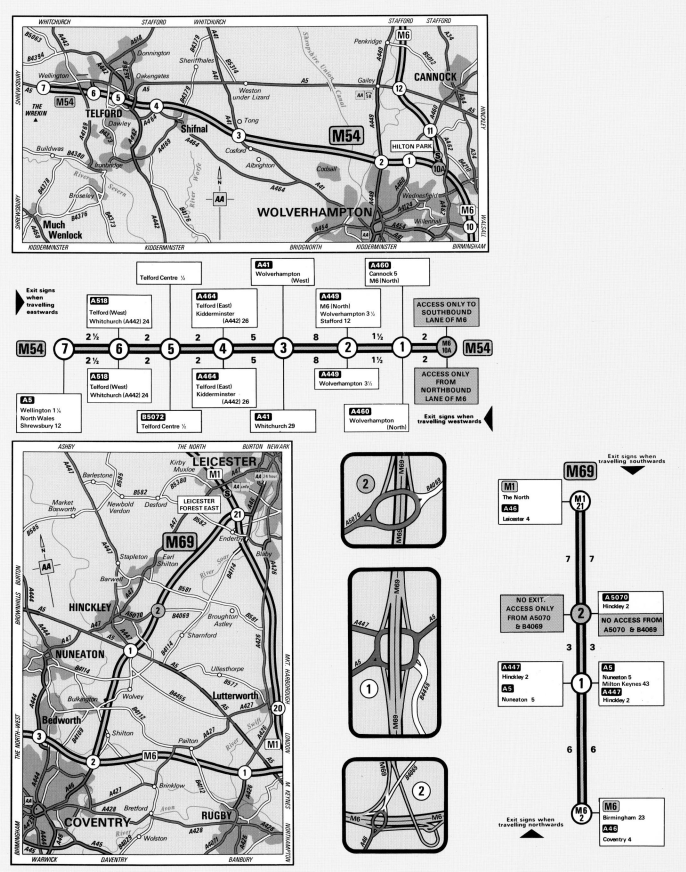

M54 map

WHITCHURCH · STAFFORD · WHITCHURCH · STAFFORD · STAFFORD

B5063 · A442 · B4394 · Donnington · Sheriffhales · A41 · B5314 · Penkridge · B5012 · A34 · A449 · Gailey · CANNOCK · A460 · A5

Wellington · A442 · Oakengates · A5 · Weston under Lizard · AA 58 · 12 · HINCKLEY

THE WREKIN · **M54** · TELFORD · A5 · B4379 · Tong · A41 · A449 · A460 · 11 · A462 · A34

Dawley · A464 · Shifnal · A41 · Cosford · 2 · 1 · HILTON PARK · S · 10A

A5223 · A464 · Albrighton · Codsall · A460 · Wednesfield · A4124 · A462 · A41 · M6

Buildwas · B4380 · Ironbridge · River Severn · A442 · A4169 · A464 · A441 · Codsall · WOLVERHAMPTON · A454 · Willenhall · 10

Broseley · B4373 · River Worfe · A442 · A454 · A41 · AA · WALSALL

Much Wenlock · A458 · B4376 · B4373 · N · AA · A442 · B4176 · A454 · BIRMINGHAM

KIDDERMINSTER · KIDDERMINSTER · BRIDGNORTH · KIDDERMINSTER · BIRMINGHAM

M54 junction strip

Exit signs when travelling eastwards ▶

| Telford Centre ½ | **A41** Wolverhampton (West) | **A460** Cannock 5 M6 (North) |

| **A518** Telford (West) Whitchurch (A442) 24 | **A464** Telford (East) Kidderminster (A442) 26 | **A449** M6 (North) Wolverhampton 3½ Stafford 12 | **ACCESS ONLY TO SOUTHBOUND LANE OF M6** |

M54 — (7) — 2½ — (6) — 2 — (5) — 2 — (4) — 5 — (3) — 8 — (2) — 1½ — (1) — 2 — M6 10A — **M54**

2½ — 2 — 2 — 5 — 8 — 1½ — 2

| **A5** Wellington 1¼ North Wales Shrewsbury 12 | **A518** Telford (West) Whitchurch (A442) 24 | **A464** Telford (East) Kidderminster (A442) 26 | **A449** Wolverhampton 3½ | **ACCESS ONLY FROM NORTHBOUND LANE OF M6** |

| **B5072** Telford Centre ½ | **A41** Whitchurch 29 | **A460** Wolverhampton (North) |

Exit signs when travelling westwards ◀

M69 map

ASHBY · THE NORTH · BURTON NEWARK

A447 · Barlestone · B585 · Kirby Muxloe · LEICESTER · M1 · A47 · AA 24 hour · AA info

Market Bosworth · Newbold Verdon · Desford · B582 · B5380 · S · LEICESTER FOREST EAST · A46 · B585

A447 · Stapleton · Earl Shilton · A47 · Enderby · 21 · **M69** · River Soar · Blaby · A426

Barwell · B581 · B581 · Broughton Astley · B581

HINCKLEY · A5070 · 2 · B4069 · Sharnford · B577

A447 · A47 · A5 · B4114 · Ullesthorpe · Lutterworth · A426

NUNEATON · B4114 · A5 · A427 · 20

Bulkington · Wolvey · B4455 · A5 · M1

Bedworth · B4112 · Shilton · Pailton · A427 · Swift · A426

3 · A428 · Brinklow · B4112 · A428

2 · M6 · COVENTRY · A428 · Bretford · Avon · RUGBY · A428 · A4071

A45 · A46 · A45 · A4029 · Wolston

WARWICK · DAVENTRY · BANBURY · NORTHAMPTON · M KEYNES · LONDON · MKT HARBOROUGH

M69 junction detail strip

Exit signs when travelling southwards ▼

M69

| **M1** The North · **A46** Leicester 4 | **M1 21** |

7 — 7

| **NO EXIT. ACCESS ONLY FROM A5070 & B4069** | **2** | **A5070** Hinckley 2 · **NO ACCESS FROM A5070 & B4069** |

3 — 3

| **A447** Hinckley 2 · **A5** Nuneaton 5 | **1** | **A5** Nuneaton 5 Milton Keynes 43 · **A447** Hinckley 2 |

6 — 6

| **M6 2** | **M6** Birmingham 23 · **A46** Coventry 4 |

Exit signs when travelling northwards ▲

Liverpool - Manchester [M62]
Manchester - Chorley [M61]
Lymm - Queensferry [M56]

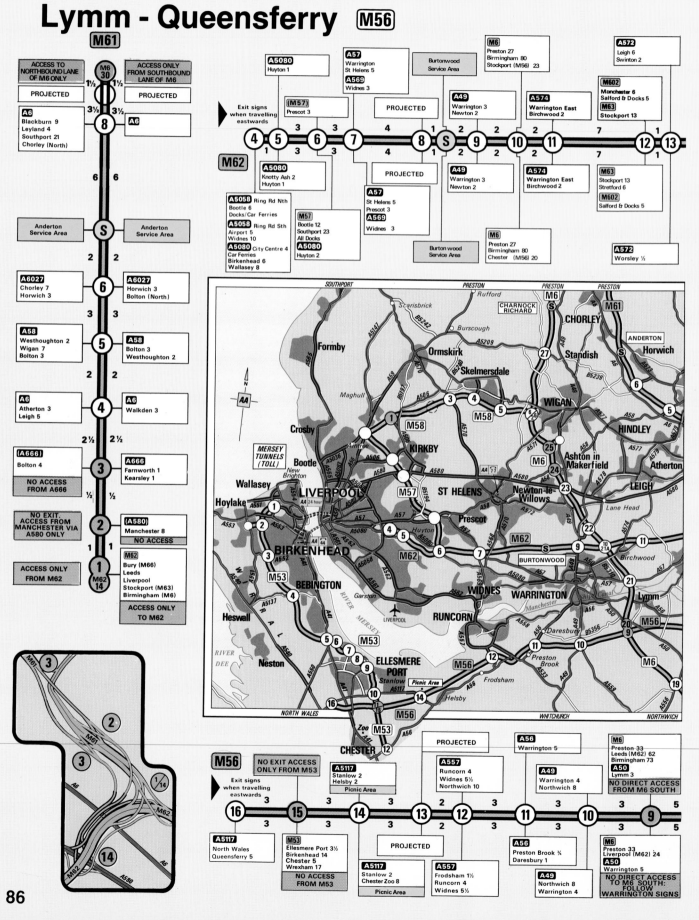

M61

ACCESS TO NORTHBOUND LANE OF M6 ONLY
PROJECTED

M6 30

ACCESS ONLY FROM SOUTHBOUND LANE OF M6
PROJECTED

A6

A6
Blackburn 9
Leyland 4
Southport 21
Chorley (North)

8

Anderton Service Area — S — Anderton Service Area

A6027
Chorley 7
Horwich 3

6

A6027
Horwich 3
Bolton (North)

A58
Westhoughton 2
Wigan 7
Bolton 3

5

A58
Bolton 3
Westhoughton 2

A6
Atherton 3
Leigh 5

4

A6
Walkden 3

(A666)
Bolton 4

3

A666
Farnworth 1
Kearsley 1

NO ACCESS FROM A666

NO EXIT. ACCESS FROM MANCHESTER VIA A580 ONLY

2

(A580)
Manchester 8
NO ACCESS

ACCESS ONLY FROM M62

1
M62 14

M62
Bury (M66)
Leeds
Liverpool
Stockport (M63)
Birmingham (M6)

ACCESS ONLY TO M62

M62

Exit signs when travelling eastwards

A5080
Huyton 1

A57
Warrington
St Helens 5

A569
Widnes 3

4 — 5 — 6 — 7 — 8 — S — 9 — 10 — 11 — 12 — 13

(M57)
Prescot 3

Burtonwood Service Area

PROJECTED

A49
Warrington 3
Newton 2

A574
Warrington East
Birchwood 2

A572
Leigh 6
Swinton 2

M602
Manchester 6
Salford & Docks 5

M63
Stockport 13

A5080
Knotty Ash 2
Huyton 1

PROJECTED

A49
Warrington 3
Newton 2

A574
Warrington East
Birchwood 2

M63
Stockport 13
Stretford 6

M602
Salford & Docks 5

A5058 Ring Rd Nth
Bootle 6
Docks/Car Ferries
A5058 Ring Rd Sth
Airport 5
Widnes 10
A5080 City Centre 4
Car Ferries
Birkenhead 6
Wallasey 8

M57
Bootle 12
Southport 23
All Docks
A5080
Huyton 2

A57
St Helens 5
Prescot 3

A569
Widnes 3

M6
Preston 27
Birmingham 80
Chester (M56) 20

A572
Worsley ½

Burton wood Service Area

M56
Exit signs when travelling eastwards

NO EXIT ACCESS ONLY FROM M53

A5117
Stanlow 2
Helsby 2
Picnic Area

PROJECTED

A557
Runcorn 4
Widnes 5½
Northwich 10

A56
Warrington 5

A49
Warrington 4
Northwich 8

M6
Preston 33
Leeds (M62) 62
Birmingham 73
A50
Lymm 3
NO DIRECT ACCESS FROM M6 SOUTH

16 — 15 — 14 — 13 — 12 — 11 — 10 — 9

A5117
North Wales
Queensferry 5

M53
Ellesmere Port 3½
Birkenhead 14
Chester 5
Wrexham 17
NO ACCESS FROM M53

PROJECTED

A5117
Stanlow 2
Chester Zoo 8
Picnic Area

A557
Frodsham 1½
Runcorn 4
Widnes 5½

A56
Preston Brook ¾
Daresbury 1

A49
Northwich 8
Warrington 4

M6
Preston 33
Liverpool (M62) 24
A50
Warrington 5
NO DIRECT ACCESS TO M6 SOUTH: FOLLOW WARRINGTON SIGNS

86

M62 Liverpool – Manchester   26 miles

Historically these two great cities have always had important transport links – the Liverpool to Manchester Railway, the Manchester Ship Canal and now the M62.

Junction 4

Liverpool has a great many hotels, restaurants and garages. Those below are closest to the junction.

WHERE TO STAY
★★**Green Park**
4–6 Greenbank Dr, Liverpool
☎051–733 3382
★**Solna**
Ullet Road, Sefton Park, Liverpool (MINO)
☎051–733 1943

WHERE TO EAT
××**Lau's**
Rankin Hall, 44 Ullet Rd, Liverpool ☎051–734 3930

GARAGES
E & A Motors
Childwall, Five Ways, Liverpool ☎051–722 2256

Junction 5

Roby and Huyton are served by this junction.

GARAGES
Avenue Service Station
Hunts Cross Av, Woolton, Liverpool ☎051–428 8483

Junction 6

The M57 branches off here linking Huyton to Aintree.

Junction 7

Widnes is accessible from this junction.

WHERE TO STAY
★★**Hillcrest**
75 Cronton Ln, Widnes
☎051–424 1616

GARAGES
Parkside
Sutton Manor, St Helens
☎(0744) 812328
Glebe
Lunts Heath Rd, Widnes (Frd)
☎051–424 5781

Burtonwood Services
(THF)
☎(0925) 51656
24hr cafeteria. Shop. Petrol. Diesel. Breakdowns. Repairs. Overnight parking for caravans £2.50. For disabled: toilets, ramp. Subway. Credit cards – shop, garage, restaurant

Junction 9

Warrington lies to the south – see M6 junctions 21 and 22 on page 68 for details.

GARAGES
A B Motors
Mill Ln, Newton-le-Willows (Dat) ☎(09252) 4411

International Auto Safety Centre
Priestley St, Warrington
☎(0925) 33308

Junction 10

The M6 links with this motorway here – see pages 64–73 for details.

Junction 11

This junction also serves Warrington.

GARAGES
Smithy
(Essendy), 32 Church Ln, Culcheth ☎(092576) 4419

Junction 12

The M63 from here skirts Manchester to reach Stockport – see page 88 – and the M602 leads into the city.

Junction 13

Here the motorway turns northwards to bypass the centre of Manchester.

GARAGES
Edge Fold
(H Owen & Co (Motor Eng)) Walkden Rd, Worsley
☎061–790 2169

M61 Manchester – Chorley   22 miles

This motorway skirts the edge of the Pennines.

Junctions 1 and 2 R

The M62 link, with Liverpool to the west and Hull to the east.

WHERE TO STAY
★**Beaucliffe**
254 Eccles Old Rd, Pendleton
☎061–789 5092

GARAGES
Gordons
Hodge Rd/Manchester Rd, Walkden (Frd)
☎061–790 2180

Junction 3 R

Bolton is to the north of this junction.

GARAGES
Century Motors
George St, Farnworth
☎(0204) 76885

Junction 4

Farmland to the south breaks up the urban sprawl around Bolton.

WHERE TO STAY
★★★**Pack Horse**
Bradshawgate, Nelson Sq, Bolton (GW) ☎(0204) 27261

GARAGES
District
Manchester Rd East, Little Hulton (BL) ☎061–790 7814

Junction 5

To the east is Bolton.

WHERE TO STAY
☆☆☆**Crest**
Beaumont Rd, Bolton (CRH)
☎(0204) 651511

WHERE TO EAT
The Lamplighter
26 Knowsley St, Bolton
☎(0204) 35175

GARAGES
Brownlow Way Garage
Brownlow Way, Bolton
☎(0204) 33300

Junction 6

Horwich is to the north.

WHERE TO STAY
INN **Coach House**
240A Warrington Rd, Lower Ince, Ince in Makerfield
☎(0942) 866330

GARAGES
B Wilkes Auto Electrics
Church St, Horwich
☎(0204) 692471

Anderton Services
(Kennings)
☎(0204) 68641
Restaurant (0700–2030. Nov – Easter) HGV cafe (south side). Shop Easter to Oct Mon – Thu 0700–2030 Fri – Sun 0700–2230. Vending machines. Petrol. Diesel. Breakdowns & Repairs 0700–2300. HGV parking (south side). Long-term/overnight parking for caravans £3. Baby-changing. For disabled: toilets, ramp. Footbridge. Road bridge. Credit cards.

Junction 8

Standing at the foot of the Pennines is Chorley.

WHERE TO STAY
★★★B **Shaw Hill Golf and Country Club**
Preston Rd, Whittle-le-Woods
☎(02572) 69221

GARAGES
Chorley Service Station
Harpers Ln, Chorley
☎(02572) 63542

M57 Huyton – Aintree

The junctions along this motorway are not numbered.

M53 Wallasey – Chester   19½ miles

This motorway crosses the Wirral to Chester.

Junction 1

Wallasey, near this junction, enjoys fine sea views.

WHERE TO STAY
★★**Grove House**
Grove Rd, Wallasey (PG)
☎051–639 3947

GARAGES
R N W
12–24 Rullerton Rd, Wallasey
☎051–342 7146

Junction 2

Hoylake is reached from this junction.

WHERE TO STAY
★★**Stanley**
Kings Gap, Hoylake (WWP)
☎051–632 3311

Junction 3

Birkenhead has grown around its ship-building yards and docks.

WHERE TO STAY
★★★B **Bowler Hat**
2 Talbot Rd, Oxton, Birkenhead (PG)
☎051–652 4931

GARAGES
Dutton-Forshaw
Park Entrance, Birkenhead (BL DJ) ☎051–647 9445

Junction 4

Port Sunlight is a town built around the Lever Brothers soap works.

WHERE TO STAY
★★★**Thornton Hall**
Neston Rd, Thornton Hough
☎051–336 3938

GARAGES
Cross Service Station
Bromborough Village Rd, Bromborough
☎051–334 2879

Junctions 5 and 6

The Liverpool University Botanic Gardens are nearby.

WHERE TO EAT
×××**Craxton Wood**
Parkgate Rd, Ledsham
☎051–339 4717

GARAGES
Dutton Forshaw
New Chester Rd, Eastham
☎051–327 1296

Junctions 7, 8 and 9

These junctions serve Ellesmere Port.

GARAGES
G Boyle
(Sports Cars) Rossmore Rd East, Ellesmere Port (Dat)
☎051–355 8354

Junction 10

The road west leads towards the Welsh border.

WHERE TO STAY
☆☆☆**Ladbroke**
(& Conferencentre) Backford Cross Rbt (LB)
☎(0244) 851551

Junction 11

Here the motorway links with the M56 – see below.

Junction 12

The motorway ends here on the edge of Chester with its many hotels and restaurants.

WHERE TO STAY
★★**Oaklands**
Hoole Rd, Chester (SNB)
☎(0244) 22156

WHERE TO EAT
××**Courtyard**
St Werburgh St, Chester
☎(0244) 21447

M56 Lymm – Queensferry   20 miles

The first part of this motorway can be found on page 88.

Junction 9 R

Access to the M6 is available from this junction.

WHERE TO STAY
★★**Lymm**
Whitbarrow Rd, Lymm (GW)
☎(092575) 2233

GARAGES
Park
(Avenue Motor Services)
Agden, Lymm
☎(092575) 2447
Howarth Motors
101 Knutsford Rd, Grappenhall, (VW Aud)
☎(0925) 65265

Junction 10

The town of Warrington is to the north.

GARAGES
Ring 'o' Bells Service Station

Northwich Rd, Lower Stretton
☎(092573) 644

Junction 11

A short way north is Daresbury, birthplace of 'Lewis Carroll'.

WHERE TO STAY
★★★**Lord Daresbury**
Chester Rd, Daresbury (GW)
☎(0925) 67331

Junction 12

Runcorn is to the north.

WHERE TO STAY
☆☆☆**Crest**
Wood Ln, Beechwood, Runcorn (CRH)
☎(0928) 714000

Junction 14

Nearby Helsby Tor (NT)

gives a superb view towards the River Mersey.

WHERE TO STAY
GH **Poplars**
130 Chester Rd, Helsby
☎(09282) 3433

Junction 15 R

This is the junction with the M53 – see above.

Junction 16

The motorway ends here and the A550 continues towards North Wales.

WHERE TO STAY
☆☆☆**Ladbroke**
(& Conferencentre), Backford Cross Rbt (LB)
☎(0244) 851551

GARAGES
Woodlands Motor Co
Chester High Rd, Burton
☎051–336 1200

M58 Aintree – Wigan

This motorway connects the M57 and the M6.

LOCAL RADIO STATIONS	Medium Wave		VHF/FM
	Metres	kHz	MHz
BBC Radio Merseyside	202	1485	95.8
BBC Radio Manchester	206	1458	95.1
IBA Radio City	194	1548	96.7
IBA Piccadilly Radio	261	1152	97.0

M62 Manchester – Huddersfield 26 miles

Between the Greater Manchester conurbation and the wool town of Huddersfield, the motorway crosses high over the Pennines.

Junction 14 R

The M61 branches off here to the north west – see page 87. Limited access to A580.

GARAGES
Gordons
Hodge Rd/Manchester Rd, Walkden (Frd)
☎061–790 2180
Dennis Fletcher & Co
Barlow St, Walkden
☎061–790 2911

Junction 15 R

Swinton and Pendlebury are residential and manufacturing areas on Manchester's north-western edge.

GARAGES
Wesley
Wesley St, Swinton
☎061–794 4916

Junction 17

The industrial town of Bury also has several interesting museums.

WHERE TO STAY
★★Racecourse
Littleton Rd, Salford
☎061–792 1420
GH Hazeldean
467 Bury New Rd, Salford
☎061–792 6667

GARAGES
Route One Ford
276 Bury New Rd, Prestwich
(Frd) ☎061–792 6161

Junction 18

The M66 heads towards Rawtenstall – see below for details.

Birch Services
(Granada)
☎061–643 0911
Restaurant. Fast food (east side). Shop. Petrol. Diesel. Breakdowns. Repairs. HGV parking. Long-term/overnight parking for caravans £3.50. Baby-changing. For disabled: toilets. Credit cards – shop, garage, restaurant. West side restaurant closed in winter – east side accessible via footbridge or tunnel.

Junction 19

More Manchester suburbs surround this junction.

WHERE TO EAT
Lancashire Fold
Kirkway, Alkrington, Middleton ☎061–643 4198

GARAGES
Coronation Service Station
Middleton Rd, Heywood
☎(0706) 69764

Junction 20

The A627(M) cuts across here on its journey from Oldham to Rochdale.

WHERE TO STAY
★★Midway
Manchester Rd, Castleton, Rochdale ☎(0706) 32881

GARAGES
Lookers
Manchester Rd, Castleton, Rochdale (DJ BL)
☎(0706) 54424
Marland
Nixon St, Rochdale
☎(0706) 31759

Junction 21

The ground starts to rise here on the edge of the Pennines.

WHERE TO STAY
★Sun
Featherstall Rd, Littleborough
☎(0706) 78957

WHERE TO EAT
×××Moorcock
Huddersfield Rd, Milnrow
☎(04577) 2659

GARAGES
Primrose Hill Service Station
Rochdale Rd, High Crompton, Shaw
☎(0706) 847438
H T H Autos
Motorama House, John St, Rochdale (Vau Opl)
☎(0706) 38491

Junction 22

As well as crossing the Pennines the road also crosses the county boundary here.

WHERE TO STAY
★★Royal
Oldham Rd, Rishworth
☎(0422) 822382

Junction 23 R

The mill town of Huddersfield is to the east in the valley of the River Colne.

Junction 24

This is another exit for Huddersfield, while the road north leads to Brighouse.

WHERE TO STAY
☆☆☆Ladbroke
(& Conferencentre) Ainley Top, Huddersfield (LB)
☎(0422) 75431
★★★George
St George's Sq, Huddersfield
☎(0484) 25444

WHERE TO EAT
Pizzeria Sole Mio
Imperial Arcade, Market St, Huddersfield
☎(0484) 42828

GARAGES
A E Crossley & Co
69 Huddersfield Rd, Elland
☎(0422) 72586
Brockholes
Southgate, Huddersfield
(Frd) ☎(0484) 29675

The motorway continues on page 90

M66 Edenfield – Manchester

The motorway begins in a valley of the Pennines above Ramsbottom and heads south into the built-up northern suburbs of Manchester. Joins M62 at Junction 18.

M67

This short stretch of motorway leads from the Manchester suburb of Denton eastwards towards Glossop. Its junctions are restricted.

M63 Salford – Stockport 14 miles

This motorway forms a west and south ring road for the Greater Manchester area.

Junction 1

For nearby restaurants, etc, see M62 junction 12 on page 87.

Junction 2

The motorway crosses the Manchester Ship Canal just south of this junction.

GARAGES
Ashmall & Parkinson
22–24–26 Wellington Rd, Eccles (BL) ☎061–789 5142

Junctions 3 to 5

The garage listed below is closest to junction 4.

GARAGES
Cascade Motors
51 Barton Rd, Urmston (Frd)
☎061–748 5531

Junction 7

From here the motorway follows the course of the River Mersey.

WHERE TO STAY
★★Simpsons
122 Withington Rd, Whalley Range ☎061–226 2235

GARAGES
Greyhound
Glebelands Rd, Sale
☎061–973 2320

Junction 8

The riverside is being developed as a recreation area nearby.

GARAGES
Cottage
375 Northendon Rd, Sale
☎061–973 0534
Warwick Service Station
142 Northendon Rd, Sale
☎061–973 7505

Junction 9 R

Northwards is the Manchester City football ground.

WHERE TO STAY
☆☆☆Post House
Palatine Rd, Northenden, (THF) ☎061–998 7090

GARAGES
Palatine
(Magnus & Pearn)
Candleford Rd, Withington
☎061–445 3514
Baguley Service Station
Altrincham Rd, Baguley
☎061–998 2683

Junctions 10 and 11

The M56 joins this motorway here.

WHERE TO STAY
☆☆☆☆Excelsior
Ringway Rd, Wythenshawe (THF) ☎061–437 5811

WHERE TO EAT
×Steak and Kebab
846 Wilmslow Rd, Didsbury
☎061–445 2552
×Cheshire Tandoori
9 The Precinct, Cheadle Hulme ☎061–485 4942

GARAGES
H & J Quick
321 Wilmslow Rd, Cheadle (Frd) ☎061–499 2141
Baguley Service Station
Altrincham Rd, Baguley
☎061–998 2683

Junction 12

South of this junction are Bramhall and Bramall Hall.

WHERE TO STAY
★★Wycliffe Villa
74 Edgeley Rd, Edgeley, Stockport ☎061–477 5395
★Acton Court
Buxton Rd, Stockport
☎061–483 6172

GARAGES
Chestergate Motors
Chestergate/King St West, Stockport ☎061–480 7395
Hollindrake
Town Hall Square, Stockport
(DJ LR Rar) ☎061–480 7966

Junction 13

For the present the motorway ends at Stockport.

WHERE TO STAY
★★★Belgrade
Dialston Ln, Stockport
☎061–483 3851

WHERE TO EAT
×Waterside
166 Stockport Rd, Romiley
☎061–430 4302

GARAGES
Stockport Road Service Station
Stockport Rd West, Bredbury
☎061–430 4938

M56 Cheadle – Altrincham 10 miles

This short stretch of motorway begins at Cheadle on a junction with the M63.

Junction 1

Bramall Hall is to the south east.

WHERE TO STAY
★★Wycliffe Villa
74 Edgeley Rd, Edgeley, Stockport ☎061–477 5395

GARAGES
H & J Quick
321 Wilmslow Rd, Cheadle
(Frd) ☎061–499 2141

Junctions 2 and 3 R

A 'B' road southwards leads to the museum and country park at Styal.

WHERE TO STAY
☆☆☆Post House
Palatine Rd, Northenden
(THF) ☎061–998 7090
GHHorizon
69 Palatine Rd, West Didsbury ☎061–445 4705

WHERE TO EAT
×Steak and Kebab
846 Wilmslow Rd, Didsbury
☎061–445 2552

GARAGES
Baguley Service Station
Altrincham Rd, Baguley
☎061–998 2683

Junction 5

Manchester Airport lies to the east.

WHERE TO STAY
☆☆☆Excelsior
Ringway Rd, Wythenshawe
(THF) ☎061–437 5811

GARAGES
Skyport Self Serve
Outwood Ln, Ringway, Manchester
☎061–499 2955

Junction 6

Residential suburbs of Manchester surround this junction.

WHERE TO STAY
★★★Valley Lodge
Altrincham Rd, Wilmslow
☎(0625) 529201

WHERE TO EAT
×Borsalino
14 The Square, Hale Barns
☎061–980 5331

GARAGES
Moores & Newton
Gresta House, Water Ln, Wilmslow (Vau Opl)
☎(0625) 527311

Junction 7

South of this junction lies Tatton Park, a fine mansion with 1,000 acres of parkland.

WHERE TO STAY
☆☆☆Ashley
Ashley Rd, Hale (GW)
☎061–928 3794
★★★HBowdon
Langham Rd, Bowdon
☎061–928 7121

WHERE TO EAT
××Evergreen
169–171 Ashley Rd, Hale
☎061–928 1222
Hale Wine Bar
106–108 Ashley Rd, Hale
☎061–928 2343

GARAGES
Bucklow
Bucklow Hill (BL)
☎(0565) 830327
Altrincham Service Station
Dunham Rd, Altrincham
☎061–928 7655

LOCAL RADIO STATIONS

	Medium Wave		VHF/FM
	Metres	kHz	MHz
BBC Radio Manchester	206	1458	95.1
BBC Radio Leeds	388	774	92.4
IBA Piccadilly Radio	261	1152	97.0
IBA Pennine Radio	235	1278	96.0

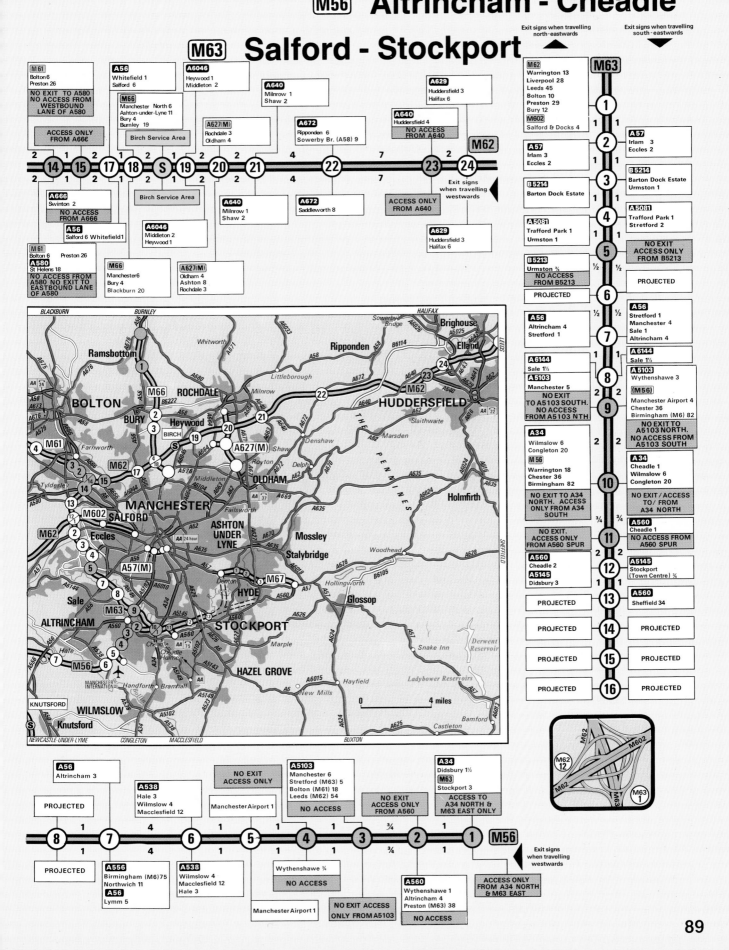

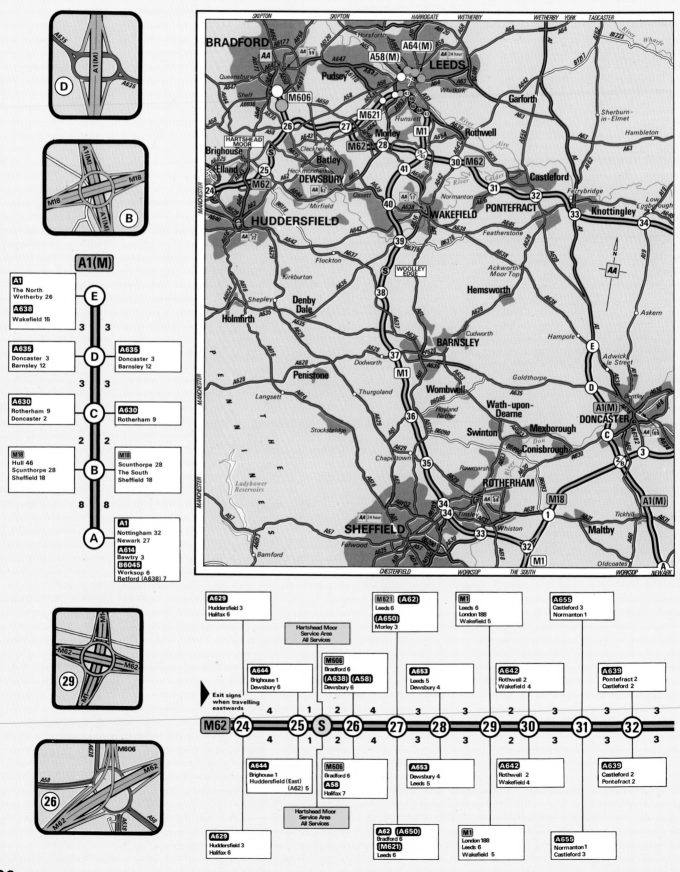

M62 Huddersfield–Castleford   28 miles

The motorway continues eastwards, serving the industrial towns of West Yorkshire.

Junction 24 see page 88

Junction 25

To the west is the mill town of Brighouse, which flourished as a canal 'port' in the 18th century.

WHERE TO STAY
☆☆☆**Ladbroke**
(& Conferencecentre) Ainley Top, Huddersfield (LB)
☎(0422) 75431
★★★**George**
St George's Sq, Huddersfield (THF) ☎(0484) 25444
GH **Cote Royd**
7 Halifax Rd, Edgerton, Huddersfield
☎(0484) 47588

WHERE TO EAT
Black Bull
Thornton Sq, Brighouse
☎(0484) 714816
GARAGES
Branch End
(R Gill) Geldard Rd, Gildersome (Suz)
☎(0532) 534311
Spring Street
(D Miller & Sons) Brighouse (Peu Tal) ☎(0484) 712222
Bailiff Bridge Service Station
Bradford Rd, Brighouse
☎(0484) 712149
Tower Auto Services
98 Church Rd, Robertown
☎(0924) 402544
Central
Huddersfield Rd, Easthorpe, Mirfield (Ren)
☎(0924) 492281

Hartshead Moor Services
(Welcome Break)
☎(0274) 876584
Restaurant. Fast food. Picnic area. HGV cafe. Shop. Vending machines. Playground. Petrol. Diesel. Liquid Petroleum Gas. Repairs. HGV parking. Long-term/overnight parking for caravans £3. Baby-changing. For disabled: toilets, ramp. Footbridge.

Junction 26

The M606 leads north to Bradford, an industrial city being modernised. An

The Leeds Crest Hotel is, in fact, six miles south-east of the city at Oulton.

Industrial Museum features the worsted textile trade and the city is also the home of the National Museum of Photography, Film and Television. Bolling Hall and Cartwright Hall can also be visited.

WHERE TO EAT
✕✕**Lillibets**
Ashfield House, 64 Leeds Rd, Liversedge
☎(0924) 404911

GARAGES
Seaguli Motor Co
Halifax Rd, Hightown, Liversedge (Lad AR)
☎(0274) 870231

M606

This short motorway takes M62 traffic into Bradford. The establishments listed below are those closest to the junctions, but others will be found in the city.

WHERE TO STAY
☆☆☆**Novotel Bradford**
Merrydale Rd, Bradford
☎(0274) 683683
☆☆☆**Stakis Norfolk Gardens**
Hall Ings, Bradford (SO)
☎(0274) 34733
★★★**Victoria**
Bridge St, Bradford (THF)
☎(0274) 728706
★**Dubrovnik**
3 Oak Av, Bradford
☎(0274) 43511
GH **Belvedere**
19 North Park Rd, Manningham, Bradford
☎(0274) 492559
GH **Maple Hill**
3 Park Dr, Heaton, Bradford
☎(0274) 44061
GH **Midway**
218 Keighley Rd, Frizinghall, Bradford ☎(0274) 42667

WHERE TO EAT
The Last Pizza Show
50 Great Horton Rd, Bradford
☎(0274) 28173
GARAGES
G D Bramall
146/148 Tong St, Bradford (Frd) ☎(0274) 681601
Townend Autos
808/810 Great Horton Rd, Bradford
☎(0274) 576285

Polar Motor Co
113/117 Manningham Ln, Bradford (Frd)
☎(0274) 305941
Park Motor Eng Co
Mansfield Rd, Oak Ln, Bradford ☎(0274) 41676

Junction 27

The M621 leads into the city of Leeds – see also M1 junction 44 on page 40. South, at Batley, there is an art gallery and a museum, while Dewsbury's parish church has some early Saxon relics.

WHERE TO STAY
GH **Alder House**
Towngate Rd, Healey Ln, Batley ☎(0924) 475540

WHERE TO EAT
✕**Tiberio**
68 Galloway Ln, Pudsey
☎(0274) 665895

GARAGES
Highway Recovery M62
Motorway House, Gelderd Rd, Birstall ☎(0924) 477474
Grahams of Dewsbury
268 Bradford Rd, Batley (Frd)
☎(0924) 472424

M621

This motorway forms an approach to Leeds from the south-western side. The junctions are not numbered and so the establishments are listed in south-to-north order.

WHERE TO STAY
★★★**Metropole**
King St, Leeds (THF)
☎(0532) 450841
☆☆☆**Ladbroke Dragonara**
Neville St, Leeds (LB)
☎(0532) 442000
★★★★**Queen's**
City Sq, Leeds
☎(0532) 431323
★★★**Merrion**
Merrion Centre, Leeds (KH)
☎(0532) 439191
★**Hartrigg**
Shire Oak Rd, Headingley
☎(0532) 751568
GH **Budapest**
14 Cardigan Rd, Headingley
☎(0532) 756637
GH **Trafford House**
18 Cardigan Rd, Headingley
☎(0532) 752034
GH **Oak Villa**
57 Cardigan Rd, Headingley
☎(0532) 758439
GH **Highfield**
79 Cardigan Rd, Headingley
☎(0532) 752193

WHERE TO EAT
✕✕✕**Gardini's Terazza**
Minerva House, 16 Greek St, Leeds ☎(0532) 432880
✕**Rules**
188 Selby Rd, Leeds
☎(0532) 604564

GARAGES
Churwell Service Station
Elland Rd, Leeds
☎(0532) 716706
G Clarke
Cross Bath Rd, Bramley, Leeds ☎(0532) 573875
Tennant Motor Service
Swinnow Ln, Leeds
☎(0532) 563411

Grandstand Service Station
247 Elland Rd, Leeds
☎(0532) 712023
International Auto Safety Centre
Armley Rd, Leeds
☎(0532) 39776
Kirkstall Service Station
Kirkstall Ln, Leeds
☎(0532) 784178
Wallace Arnold Sales and Service
123 Hurslet Rd, Leeds (Vau Opl) ☎(0532) 439911
Ringways
Whitehall Rd, Leeds (Frd)
☎(0532) 634222
KP Motor Cycles
No 1 The Calls, Leeds (Hon Yam Suz) ☎(0532) 460705
Brown & Rose
Lavender Walk, Leeds
☎(0532) 482019
A G Wilson
Regent St, Leeds (BL LR Rar)
☎(0532) 438201

Junction 28

M62 – a junction for Leeds and Dewsbury.

WHERE TO STAY
☆☆☆☆**Post House**
Queen's Dr, Ossett, Wakefield (THF)
☎(0924) 276388

GARAGES
Heybeck
Leeds Rd, Woodkirk
☎(0924) 472660
Rocar
Aldams Rd, Dewsbury (BL)
☎(0924) 465652

LOCAL RADIO STATIONS			
	Medium Wave		VHF/FM
	Metres	kHz	MHz
BBC Radio Leeds	388	774	92.4
IBA Radio Aire	362	828	94.6

West Riding House towers over the rest of the buildings in Leeds' Commercial Street.

Junction 29

This is the M62 junction with the M1 – see pages 34–41 for details.

Junction 30

Wakefield, capital of Yorkshire's woollen industry for 700 years until the coming of the factory age, is to the south. The Art Gallery contains modern paintings and sculpture and there is also a museum.

WHERE TO STAY
☆☆☆**Leeds Crest**
The Grove, Oulton (CRH)
☎(0532) 826201
★★★**Swallow**
Queens St, Wakefield (SW)
☎(0924) 372111

WHERE TO EAT
Venus Restaurant
51 Westgate, Wakefield
☎(0924) 75378

GARAGES
Coopers
Aberford Rd, Oulton (BL)
☎(0532) 822382
Goodycross
Goodycross Ln, Swillington
☎(0532) 871963
Appleyard
Ings Rd, Wakefield (BL)
☎(0924) 70100
Glanfield Lawrence
68 Ings Rd, Wakefield (Vau Opl) ☎(0924) 372812
Wensley's
68 Ings Rd, Wakefield (VW Aud) ☎(0924) 375588

Junction 31

The colliery town of Castleford was the birthplace, in 1898, of Henry Moore, the sculptor.

WHERE TO STAY
★★★**Stoneleigh**
Doncaster Rd, Wakefield
☎(0924) 369461

GARAGES
Arnold G Wilson
Doncaster Rd, Wakefield (LR BL DJ Rar)
☎(0924) 377261

Junction 32

Pontefract is to the south.

GARAGES
New Quarry Service Station
Wakefield Rd, Pontefract
☎(0977) 702468
Martin
(Pontefract) Oxford St, Castleford (Frd)
☎(0977) 558301
Seniors Motorcycles
11 and 13 High St, Ferrybridge (Hon Ves Lam)
☎(0977) 82581

For information on the A1(M) see page 92.

For information on the M1 see pages 34–41.

M62 Pontefract–Hull   28 miles

We leave industrial areas behind to head eastwards across low ground towards the Humber estuary.

Junction 33

Knottingley is to the north of this junction. The River Aire and the Aire and Calder Canal run along its northern boundary.

WHERE TO STAY
★★★ ♨H **Wentbridge House**
Wentbridge
☎(0977) 620444

WHERE TO EAT
Bay Horse
Fairburn
☎(0977) 85126/82371

M18 Rotherham–Goole   27 miles

This motorway links industrial South Yorkshire to the Humberside docks, crossing flat, open country on the way.

Junction 1

After leaving the M1 east of Sheffield, this first junction is for Rotherham.

WHERE TO STAY
★**Elton**
Main St, Bramley
☎(0709) 545681

GARAGES
E Butler Auto Sales
(E Butler & Sons) Blyth Rd, Maltby (BL)
☎(0709) 813161

Junction 2

A1(M) crosses here – see this page, right, for details.

Junction 3

Doncaster is known for its racecourse, railway works and also its butterscotch.

WHERE TO STAY
★★★**Punch's**
Bawtry Rd, Bessacar, Doncaster (EH)
☎(0302) 535235
★★★**Earl of Doncaster**
Bennetthorpe, Doncaster
☎(0302) 61371

Junction 4

Access from this junction is to the north side of Doncaster. The racecourse, where the St Leger is run every September, is on the eastern side of the town.

Junction 5

The M180 leads off to the east towards Humberside Airport and Grimsby – see below for details.

Junction 6

Beyond the small town of Thorne, on the Stainford and Keadby Canal, moorland stretches away to the east.

WHERE TO STAY
★★**Belmont**
Horsefair Green, Thorne
☎(0405) 812320

GARAGES
Adams Fairfield
16 Fieldside, Thorne (Frd, BL)
☎(0405) 812345

Junction 7

The M18 ends here at its junction with the M62 – see above for details.

GARAGES
Seniors Motorcycles
11 & 13 High St, Ferrybridge (Hon Ves Lam)
☎(0977) 82581
Motorama
Hill Top, Knottingley
☎(0977) 83873
New Quarry Service Station
Wakefield Rd, Pontefract
☎(0977) 702468

Junction 34

Selby, to the north-east, although only a small market town, has a fine church, once part of Selby Abbey.

★★B **Regent**
Regent Sq, Doncaster
☎(0302) 64180
★★★**Danum**
High St, Doncaster (SW)
☎(0302) 62261
★★**Mount Pleasant**
Rossington ☎(0302) 868696

WHERE TO EAT
Vintage Steak Bar
Cleveland St, Doncaster
☎(0302) 64786
Ristorante Il Fiore in Legards
50–51 High St, Doncaster
☎(0302) 23287
Bacchus
44 Hallgate, Doncaster
☎(0302) 20232
Pizzeria San Remo
8 Netherhall Rd, Doncaster
☎(0302) 60501

GARAGES
E W Jackson & Son
1 Church Way, Doncaster
(BL) ☎(0302) 21541
Kennings
York Rd, Doncaster (BL)
☎(0302) 780780
Ringways
York Rd, Doncaster (Frd)
☎(0302) 785211

GARAGES
Station
(L R Carroll & Son) Whitley Bridge ☎(0977) 661256

Junction 35

The M18 joins from the south – see below for details. To the north west stands Carlton Towers.

Junction 36

Goole is a busy port 50 miles inland from the North Sea which depends on the Yorkshire Ouse and the docks of the Aire and Calder Navigation.

WHERE TO STAY
★★**Clifton**
1 Clifton Gardens, Boothferry Rd, Goole ☎(0405) 61336

GARAGES
Glews
Rawcliffe Rd, Goole (Tal Peu)
☎(0405) 4525
Kennings
Rawcliffe Rd, Goole
☎(0405) 3444

Junction 37

The motorway crosses the wide River Ouse to reach this junction by Howden.

WHERE TO STAY
★★**Bowmans**
Bridgegate, Howden
☎(0430) 30805
★**Wellington**
31 Bridgegate, Howden
☎(0430) 30258

Junction 38

The motorway ends here and the A63 completes the journey to Hull.

LOCAL RADIO STATIONS

	Medium Wave		VHF/FM
	Metres	kHz	MHz
BBC Radio Sheffield	290	1035	97.4
BBC Radio Humberside	202	1485	96.9
IBA Radio Hallam			
Rotherham area	194	1548	95.9
IBA Viking Radio	258	1161	NA

A1 (M)

Every so often on its journey northwards, the A1 takes on motorway status. This short stretch bypasses Doncaster.

Junction A

By Whitewater Common the A1 becomes motorway.

WHERE TO STAY
★★**Fourways**
Blyth (090976) 235
★★★**Ye Olde Bell**
Barnby Moor (THF)
☎(0777) 705121
★★★**Crown**
High St, Bawtry (AHT)
☎(0302) 710341

WHERE TO EAT
Maigret's Wine Bar
High St, Bawtry
☎(0302) 711057

GARAGES
Holmgarth Motor Company
Bawtry Rd, Blyth
☎(0777) 706293

Junction B

The M18 junction – see this page, left, for details.

WHERE TO STAY
★★★**Punch's**
Bawtry Rd, Bessacarr, Doncaster (EH)
☎(0302) 535235

Junction C

Doncaster is to the east – see also M18 junctions 3 and 4 on this page. South-west is Conisborough with its 12th-century castle.

WHERE TO STAY
★★★**Danum**
High St, Doncaster (SW)
☎(0302) 62261
★★B **Regent**
Regent Sq, Doncaster
☎(0302) 64180
★★★**Earl of Doncaster**
Bennetthorpe, Doncaster
☎(0302) 61371

WHERE TO EAT
Vintage Steak Bar
Cleveland St, Doncaster
☎(0302) 64786
Ristorante Il Fiore in Legards
50–51 High St, Doncaster
☎(0302) 23287
Bacchus
44 Hallgate, Doncaster
☎(0302) 20232
Pizzeria San Remo
8 Netherhall Rd, Doncaster
☎(0302) 60501

GARAGES
Station Supreme
High Rd, Warmsworth (Hon)
☎(0302) 855506
E W Jackson & Son
1 Church Way, Doncaster
(BL) ☎(0302) 21541

Junction D

The Cusworth Country Park is just to the south.

GARAGES
Green Lane
(S Plumb & Son) Scawthorpe, Doncaster (Maz)
☎(0302) 723523
Kennings
York Rd, Doncaster (BL)
☎(0302) 780780
Ringways
York Rd, Doncaster (Frd)
☎(0302) 785211

Junction E

The motorway restrictions end here and the A1 continues northwards.

WHERE TO STAY
☆☆☆**TraveLodge**
Barnsdale Bar (THF)
☎(0977) 620711

WHERE TO EAT
××**Hampole Priory**
Hampole ☎(0302) 723740

LOCAL RADIO STATIONS

	Medium Wave		VHF/FM
	Metres	kHz	MHz
BBC Radio Sheffield	290	1035	97.4
IBA Radio Hallam			
Rotherham area	194	1548	95.9

M180 Thorne–Brigg   26 miles

South Humberside is crossed by this motorway, passing Scunthorpe on its way towards Grimsby, Immingham Docks and the southern approach to the Humber Bridge.

Junction 1

Just off the M18, this junction has Thorne to the north.

GARAGES
Carr
(F Cross & Sons) Thorne Rd, Hatfield (Toy)
☎(0302) 840348
Green Tree
Tudworth Rd, Hatfield
☎(0302) 840488

Junction 2

The developing National Transport Museum at Sandtoft includes many historic buses. South, Epworth's old Rectory is open – birthplace of John and Charles Wesley.

LOCAL RADIO STATIONS

	Medium Wave		VHF/FM
	Metres	kHz	MHz
BBC Radio Humberside	202	1485	96.9
IBA Viking Radio	258	1161	NA

Junction 3

The M181 leads north into Scunthorpe, an industrial town at the heart of the largest ironstone beds in Europe.

WHERE TO STAY
★★**Royal**
Doncaster Rd, Scunthorpe (AHT) ☎(0724) 868181
★★**Wortley**
Rowland Rd, Scunthorpe
☎(0724) 842223

WHERE TO EAT
×**Town House**
62 Mary St, Scunthorpe
☎(0724) 863692
Amina Indian Restaurant
25–27 Cole St, Scunthorpe
☎(0724) 840823

GARAGES
Turners
(Berkeley Services)
Doncaster Rd, Scunthorpe
(Col) ☎(0724) 860212
Kennings
Normanby Rd, Scunthorpe
(BL LR Rar) ☎(0724) 856551
Central
101 Ashby High St,
Scunthorpe (Cit)
☎(0724) 842909
Leys Farm
Park Av, Bottesford,
Scunthorpe ☎(0724) 868731

Junction 4

To the north of Scunthorpe is Normanby Hall. The 350-acre park includes a Countryside Interpretation Centre, a Blacksmith's Shop and a potter at work.

WHERE TO EAT
PS **Raventhorpe Farm**
N side of junction A18/B1398.
OS112 SE9307

GARAGES
Lockwoods Scawby
☎(0652) 54310

Junction 5

South-west of this final junction is the Wrawby Postmill, the last surviving example of its type in the north of England. Northwards is Elsham Hall Country Park. The A15 north leads to the Humber Bridge (toll).

GARAGES
John Morris Motors
Wootton (Vlo) ☎(04695) 512

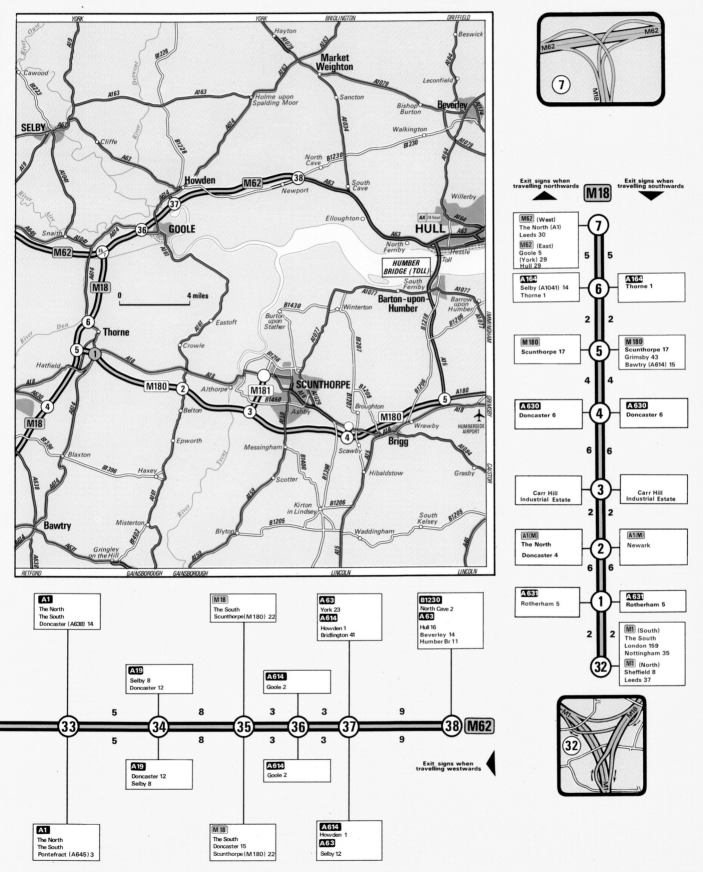

M18

Exit signs when travelling northwards / Exit signs when travelling southwards

Northbound	Jct	Southbound
M62 (West) The North (A1) Leeds 30 / **M62** (East) Goole 5 (York) 29 Hull 29	7	
A164 Selby (A1041) 14 Thorne 1	6	**A164** Thorne 1
M180 Scunthorpe 17	5	**M180** Scunthorpe 17 Grimsby 43 Bawtry (A614) 15
A630 Doncaster 6	4	**A630** Doncaster 6
Carr Hill Industrial Estate	3	Carr Hill Industrial Estate
A1(M) The North Doncaster 4	2	**A1(M)** Newark
A631 Rotherham 5	1	**A631** Rotherham 5
	32	**M1** (South) The South London 159 Nottingham 35 / **M1** (North) Sheffield 8 Leeds 37

Distances between junctions: 7–6: 5, 6–5: 2, 5–4: 4, 4–3: 6, 3–2: 2, 2–1: 6, 1–32: 2

M62

Jct 33	34	35	36	37	38

Distances: 33–34: 5, 34–35: 8, 35–36: 3, 36–37: 3, 37–38: 9

Northbound boxes:
- **A1** The North The South Doncaster (A638) 14
- **A19** Selby 8 Doncaster 12
- **M18** The South Scunthorpe (M180) 22
- **A63** York 23 **A614** Howden 1 Bridlington 41
- **A614** Goole 2
- **B1230** North Cave 2 **A63** Hull 16 Beverley 14 Humber Br 11

Southbound/westbound boxes:
- **A1** The North The South Pontefract (A645) 3
- **A19** Doncaster 12 Selby 8
- **M18** The South Doncaster 15 Scunthorpe (M180) 22
- **A614** Goole 2
- **A614** Howden 1 **A63** Selby 12

Exit signs when travelling westwards

Leyland - Colne [M65]

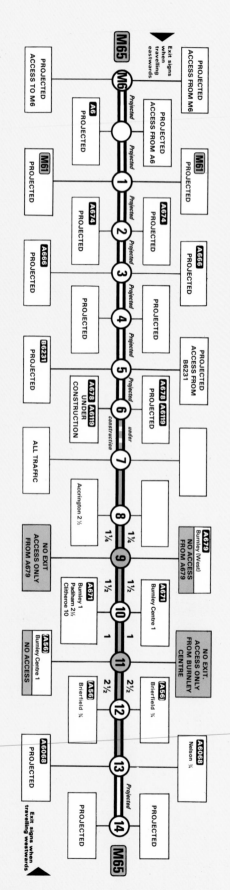

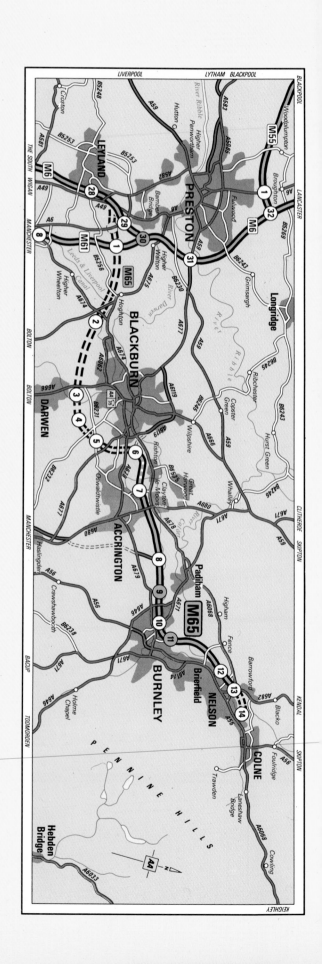

Junction 7

The section to the west of this junction is under construction and is due to be completed during the summer of 1985. Accrington, to the south, is known for its brick industry as well as textiles. The Haworth Art Gallery contains a large collection of Tiffany glass.

WHERE TO EAT
×××**Foxfields**
Whalley Rd, Billington
☎(025482) 2556

GARAGES
Reservoir Service Station
(Autotune) Blackburn Rd,
Rishton ☎(0254) 56962
Leyland Street
Leyland St, Accrington
☎(0254) 32424

Junction 8

At Padiham, to the north, is Gawthorpe Hall, an early 17th-century house with a minstrels' gallery. The Kay Shuttleworth collection of lace and embroidery is also on display.

WHERE TO EAT
××**Belvedere**
Read ☎(0282) 72170

Once witches were said to cast their spells in the shadow of Pendle Hill. In recent times a happier association has been the custom of rolling Easter eggs down its 1,831 feet.

Junction 9 R

Hameldon Hill rises to 1342 feet to the south.

Junctions 10 and 11 R

At Burnley heavy industry now dominates what was once the largest cotton-weaving centre in the world. Towneley Hall is a 14th-century house with later modifications. It contains some period rooms and an art gallery and museum.

WHERE TO STAY
★★★**Keirby**
Keirby Walk, Burnley
☎(0282) 27611
★★**Rosehill House**
Rosehill Av, Burnley (BW)
☎(0282) 53931

WHERE TO EAT
Smackwater Jacks
Ormerod St, Burnley
☎(0282) 21290

GARAGES
Holden & Hartley
Accrington Rd, Burnley (Vau
Opl) ☎(0282) 27321
Hebden Bros
Todmorden Rd, Burnley (BL
DJ) ☎(0282) 36131
Pendle
Church St, Burnley (Dat)
☎(0282) 38738

Vale
Abinger St, Burnley
☎(0282) 26250

Junction 12

To the west is Pendle Hill, well known for its witchcraft stories. Nelson developed as a textile centre and took its name from the Nelson Inn around which it grew.

GARAGES
Ratcliffe & Thornton Bros
Manchester Rd, Nelson (BL
DJ LR) ☎(0282) 66771

Junction 13

The motorway ends just short of Colne on the edge of the Pennines. In the town is an interesting museum on the British in India up to 1947, with dioramas, a model railway and many other items. The Forest of Trawden and Keighley Moor stretch away to the east.

WHERE TO STAY
★★**Great Marsden**
Barkerhouse Rd, Nelson
☎(0282) 64749

WHERE TO EAT
××**Barley Mow**
Barley ☎(0282) 64293

GARAGES
Popular
Nora St, Barrowford
☎(0282) 64759

M65 Leyland–Colne   6¼ miles open

This motorway, when complete, will link a number of industrial towns west of the Pennines.

LOCAL RADIO STATIONS			
	Medium Wave		**VHF/FM**
	Metres	*kHz*	*MHz*
BBC Radio Lancashire	352	855	96.4

Gawthorpe Hall is a splendid Jacobean house on the banks of the River Calder. Textile craft courses and exhibitions are among its attractions.

Kincardine Bridge - Dunblane M9
Glasgow - Bishopton M8
Dennyloanhead - Kincardine Bridge M876

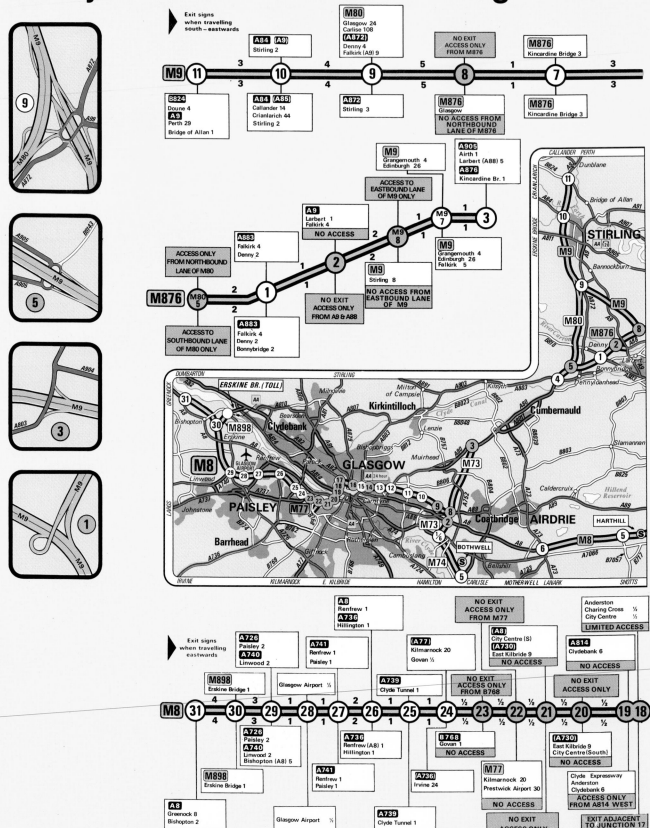

Exit signs when travelling south – eastwards

M80 Glasgow 24 / Carlise 108 / (A872) / Denny 4 / Falkirk (A9) 9

A84 (A9) Stirling 2

NO EXIT ACCESS ONLY FROM M876

M876 Kincardine Bridge 3

M9 (11) — 3 — (10) — 4 — (9) — 5 — (8) — 1 — (7) — 3

B824 Doune 4 / **A9** Perth 29 / Bridge of Allan 1

A84 (A85) Callander 14 / Crianlarich 44 / Stirling 2

A872 Stirling 3

M876 Glasgow / NO ACCESS FROM NORTHBOUND LANE OF M876

M876 Kincardine Bridge 3

M9 Grangemouth 4 / Edinburgh 26

ACCESS TO EASTBOUND LANE OF M9 ONLY

A905 Airth 1 / Larbert (A88) 5 / **A876** Kincardine Br. 1

ACCESS ONLY FROM NORTHBOUND LANE OF M80

A883 Falkirk 4 / Denny 2

A9 Larbert 1 / Falkirk 4 / NO ACCESS

M9 Grangemouth 4 / Edinburgh 26 / Falkirk 5

M876 **M80** (5) — 2 — (1) — 1 — (2) — 2 — (8) — 1 — (7) — 1 — (3)

ACCESS TO SOUTHBOUND LANE OF M80 ONLY

A883 Falkirk 4 / Denny 2 / Bonnybridge 2

NO EXIT ACCESS ONLY FROM A9 & A88

M9 Stirling 8

M9 NO ACCESS FROM EASTBOUND LANE OF M9

Exit signs when travelling eastwards

A8 Renfrew 1 / **A736** Hillington 1

NO EXIT ACCESS ONLY FROM M77

Anderston / Charing Cross ½ / City Centre ½ / LIMITED ACCESS

A726 Paisley 2 / **A740** Linwood 2

A741 Renfrew 1 / Paisley 1

(A77) Kilmarnock 20 / Govan ½

(A8) City Centre (S) / (A730) East Kilbride 9 / NO ACCESS

A814 Clydebank 6 / NO ACCESS

M898 Erskine Bridge 1

Glasgow Airport ½

A739 Clyde Tunnel 1

NO EXIT ACCESS ONLY FROM B768

NO EXIT ACCESS ONLY

M8 (31) — 4 — (30) — 3 — (29) — 1 — (28) — 2 — (27) — 1 — (26) — 2 — (25) — 1 — (24) — ½ — (23) — ½ — (22) — ½ — (21) — ½ — (20) — ½ — (19) — (18)

A726 Paisley 2 / **A740** Linwood 2 / Bishopton (A8) 5

A736 Renfrew (A8) 1 / Hillington 1

B768 Govan 1 / NO ACCESS

(A730) East Kilbride 9 / City Centre (South) / NO ACCESS

M898 Erskine Bridge 1

A741 Renfrew 1 / Paisley 1

(A736) Irvine 24

M77 Kilmarnock 20 / Prestwick Airport 30 / NO ACCESS

Clyde Expressway Anderston / Clydebank 6 / ACCESS ONLY FROM A814 WEST

A8 Greenock 8 / Bishopton 2

Glasgow Airport ½

A739 Clyde Tunnel 1

NO EXIT ACCESS ONLY

EXIT ADJACENT TO JUNCTION 17 / LIMITED ACCESS

96

M9 Kincardine Bridge– Dunblane   13 miles

The first part of this motorway is detailed on page 98. Here it continues on the southern side of the River Forth.

Junction 7

The M876 branches off northwards towards the Kincardine Bridge over the Firth of Forth.

WHERE TO EAT
PS **Kincardine Bridge**
Near S access on A88
OS65 NS9185

Junction 8 R

The M 876 heads south-west towards Glasgow.

Junction 9

To the north is Bannockburn, scene of the momentous Battle in 1314 when Robert the Bruce defeated Edward II's army and secured Scotland's independence. An equestrian statue of Bruce commemorates the event.

Junction 10

Stirling is a historic town,

home of the Stewart kings for centuries, and now a popular tourist attraction with many places of interest. North-west is Scotland's Safari Park at Blair Drummond.

WHERE TO STAY
★★★**Golden Lion**
King St, Stirling
☎(0786) 5351
★★**Stakis Station**
56 Murray Pl, Stirling (SO)
☎(0786) 2017
★**Kings Gate**
5 King St, Stirling
☎(0786) 3944

WHERE TO EAT
××**Heritage**
16 Allan Park, Stirling
☎(0786) 3660
××**Blairlogie House**
Blairlogie ☎(0259) 61441
Riverway Restaurant
Kildean, Stirling
☎(0786) 5734
Qismat
37 Friar's St, Stirling
☎(0786) 63075

GARAGES
Mogil Motors
Drip Rd and Union St, Stirling
(Frd) ☎(0786) 4891
Graham and Morton
Kerse Rd, Stirling (AR Ren)

Maz Col Dhu)
☎(0786) 2212

Junction 11

The motorway ends here, just south of Dunblane, a small but ancient city which is dominated by its 13th-century cathedral. South-east is the 220ft Wallace Monument with audio-visual presentation, while to the west is Doune with its restored 14th-century castle and its motor museum.

WHERE TO STAY
★★★**Royal**
Henderson St, Bridge of Allan (BW) ☎(0786) 832284
★★★**Stakis Dunblane Hydro**
Dunblane (SO)
☎(0786) 822551

WHERE TO EAT
Fourways Restaurant
Main North Rd, Dunblane
☎(0786) 822098

GARAGES
Stirling Road
(Doune Motors) Stirling Rd, Doune (078684) 361

M876 Dennyloanhead– Kincardine Bridge   7 miles

This short motorway cuts a corner for travellers to and from the Kincardine Bridge over the Forth.

Junction 1

A little way south is the course of the Antonine Wall, beside which is Rough Castle, one of Britain's most remarkable Roman military sites.

WHERE TO STAY
☆☆**Stakis Park**
Arnot Hill, Camelon Rd, Falkirk (SO) ☎(0324) 28331

WHERE TO EAT
❀×**Pierre's**
140 Graham's Rd, Falkirk
☎(0324) 35843

Junction 2 R

Falkirk is a historic city

Cladhan
Kemper Ave, Falkirk
☎(0324) 27421

GARAGES
Millars
Callendar Rd, Falkirk (Frd)
☎(0324) 21511
Northern
(Johnston Stewart & Sons)
Bainsford, Falkirk (DJ BL)
☎(0324) 22584
Square Deal Motors
Ladysmill, Falkirk (Toy)
☎(0324) 35935

LOCAL RADIO STATIONS

	Medium Wave		VHF/FM
	Metres	kHz	MHz
IBA Radio Forth	194	1548	96.8

with iron work its main industry for some 200 years. Cannons for Nelson's *Victory* were manufactured here.

M9 Junction

The motorway joins the M9 between its junctions 8 and 7 before branching off again.

Junction 3

The motorway ends just short of the Kincardine Bridge.

WHERE TO EAT
PS **Kincardine Bridge**
Near S access on A88
OS65 NS9185

GARAGES
Kincardine Motors
Feregait, Kincardine-on-Forth
(Frd) ☎(0259) 30230

M8 Glasgow–Bishopton   15½ miles

A string of motorway junctions skirt the centre of Glasgow, then the road heads west to end on the south bank of the Clyde. The preceding part of this motorway can be found on page 98.

Junctions 18 and 19 R

One of Glasgow's main shopping streets, Argyll Street, is adjacent to this junction. There are many hotels, restaurants and garages in Glasgow – space allows us only to list those nearest to the junctions.

WHERE TO STAY
★★★★**Albany**
Bothwell St, Glasgow (THF)
☎041–248 2656
☆☆☆**Holiday Inn Glasgow**
Argyle St, Anderston (CHI)
☎041–226 5577
☆☆**Crest Glasgow-City**
Argyle St, Anderston (CRH)
☎041–248 2355
GH **Smith's**
963 Sauchiehall St, Glasgow
☎041–339 7674

WHERE TO EAT
×××**Ambassador**
19/20 Blythswood Sq,
Glasgow ☎041–221 2034
⛟××**Buttery**
652 Argyle St, Glasgow
☎041–221 8188
×**Trattoria Toscana**
47 Robertson St, Glasgow
☎041–221 4330
Belfry
652 Argyle St, Glasgow
☎041–221 0630
Epicures Bistro
46 West Nule St, Glasgow
☎041–221 7488

GARAGES
Nicol Recovery
142 Main St, Bridgeton
☎041–554 1025
MacHarg Rennie & Lindsay
188 Castlebank St, Glasgow
(BL LR)
Scotstoun Motors
18 Harland St, Glasgow
☎041–954 4666

Junctions 20 and 21 R

The motorway crosses Kingston Bridge to the south bank of the Clyde. The short M77 leads south to Pollock Country Park, which now contains the fabulous Burrell Collection. Haggs Castle is nearby.

WHERE TO STAY
★★**Ewington**
Queen's Drive, Queens Park
☎041–423 1152
★★★**Tinto Firs Thistle**
470 Kilmarnock Rd, Glasgow
(TS) ☎041–637 2353
GH **Marie Stuart**
46–48 Queen Mary Ave,
Cathcart ☎041–423 6363

WHERE TO EAT
××**Kensingtons**
164 Darnley St, Pollokshields
☎041–424 3662

GARAGES
Blairhall Motors
70 Stanley St, Glasgow
☎041–429 1121
Portman
31 Portman St, Glasgow
☎041–429 6701
Wylies
370 Pollokshaws Rd,
Glasgow (Frd)
☎041–423 6644

LOCAL RADIO STATIONS

	Medium Wave		VHF/FM
	Metres	kHz	MHz
IBA Radio Clyde	261	1152	95.1

Junction 23 R

Bellahouston Park is beside this junction, the site of the 1938 Empire Exhibition.

WHERE TO STAY
★★★**Bellahouston**
517 Paisley Rd West,
Glasgow (SW)
☎041–427 3146
★★**Sherbrooke**
11 Sherbrooke Ave,
Pollokshields
☎041–427 4227
GH **Linwood**
356 Albert Dr, Pollokshields
☎041–427 1642

WHERE TO EAT
××**Massimo's**
465 Clarkston Rd, Muirend
☎041–637 8568

GARAGES
Park Auto Co
69 Dumbreck Rd, Glasgow
(BL) ☎041–427 2206
Carlaw Cars
222 Nether Auldhouse Rd,
Pollokshaws (DJ BL)
☎041–649 4585

Junction 24

Glasgow Rangers football ground at Ibrox Park is just to the north.

GARAGES
Hydepark Motor Eng
49 Shieldhall Rd, Glasgow
☎041–440 1655

Junction 25

Moving away from central Glasgow now, the motorway enters an industrial area.

WHERE TO STAY
★★**Wickets**
52–54 Fortrose St, Partickhill
☎041–334 9334

Junction 26

Once the county town of the former Renfrewshire, Renfrew has now become part of the industrial conurbation which sprawls along the River Clyde here.

WHERE TO STAY
☆☆**Dean Park**
91 Glasgow Rd, Renfrew
☎041–886 3771
☆☆☆**Normandy**
Inchinnan Rd, Renfrew
☎041–886 4100

GARAGES
Mossview Motors
450 Shieldhall Rd, Glasgow
(Dat) ☎041–445 3939
Glebe Coachworks
6 Glebe St, Renfrew
☎041–885 1818

Junction 27

Paisley, to the south, is famous for the Paisley shawl. A priceless collection of these garments survives in the town's museum.

WHERE TO STAY
★★**Rockfield**
12 Renfrew Rd, Paisley (ICA)
☎041–889 6182
★★★**Glynhill**
Paisley Rd, Renfrew
☎041–886 5555
★★**Ardgowan**
Blackhall St, Lonend, Paisley
☎041–887 2196
GH **Broadstones**
17 High Calside, Paisley
☎041–889 4055

WHERE TO EAT
Cardosi's
46 Causeyside St, Paisley
☎041–889 5339

GARAGES
J D Harvey Motors
Porterfield Rd, Renfrew
☎041–886 4009
Longcroft
(D W & H Ferguson)
Porterfield Rd, Renfrew (Vau
Opl) ☎041–886 2777
Moorpark Service Station
132 Paisley Rd, Renfrew
☎041–886 4009

Junction 28

This is the junction for Glasgow Airport, right next to the motorway.

WHERE TO STAY
★★★★**Excelsior**
Abbotsinch (THF)
☎041–887 1212

Junction 29

Glasgow Airport is still alongside the motorway.

WHERE TO STAY
★★**Golden Pheasant**
Moss Rd, Linwood (ICA)
☎(0505) 21266

GARAGES
Hunters
80 Clark St, Paisley
☎041–889 2414
Elderslie Service Station
162–192 Main Rd, Elderslie
(Sko) ☎(0505) 22705
Wallace Service Station
Main Rd, Elderslie
☎(0505) 22374

Junction 30

The M898 leads to Erskine and the Erskine Bridge across the Clyde. On the north side is an area of hills, lochs and forest.

WHERE TO STAY
☆☆☆**Crest**
Erskine Bridge, North Barr,
Inchinnan (CRH)
☎041–812 0123

GARAGES
Clydeholm
(R D Laidlaw Co) Napier St,
Clydebank ☎041–952 1325

Junction 31

The motorway ends on the south bank of the Clyde, with Dumbarton Castle prominent on the opposite bank.

WHERE TO STAY
★★★ ⬛BL **Gleddoch House**
Langbank ☎(047554) 711

GARAGES
Holmpark
(T D & C M Elvidge) Greenock
Rd, Bishopton
☎(05086) 2511
Kilbarchan Service Station & Eng Co
Milliken Park, Kilbarchan
☎041–542 2342

M8 Edinburgh–Glasgow   38 miles

This motorway links Scotland's two major cities, passing through some attractive countryside on the way.

Edinburgh Castle stands high on its rocky perch above the gardens of Princes Street.

Junction 2

Edinburgh Airport is just to the east, with the Royal Highland Showground adjacent.

WHERE TO STAY
FH Mr and Mrs Pollock, **Easter Norton Farm**, Newbridge ☎031–333 1279

M9 Edinburgh–Kincardine Bridge   15 miles

From Scotland's capital, this motorway follows the Firth of Forth between the Forth and Kincardine bridges.

Junction 1 R

The M8 continues south and westwards for Glasgow. To the north is the massive Forth Road Bridge (toll) and its more famous neighbour, the railway bridge.

WHERE TO STAY
☆☆☆**Forth Bridges Moat House**
Forth Road Bridge, Queensferry (South) (QM)
☎031–331 1199

WHERE TO EAT
✕✕**Hawes Inn**
Queensferry (South)
☎031–331 1990

GARAGES
Hawes Garage
(G A McTeague) Queensferry (South) ☎031–331 1796

Junction 2 R

On the bank of the Forth is Hopetoun House, a splendid building with a commanding view. It houses a famous art collection.

LOCAL RADIO STATIONS

	Medium Wave		VHF/FM
	Metres	kHz	MHz
IBA Radio Forth	194	1548	96.8
IBA Radio Clyde	261	1152	95.1

Junction 3 R

The ruins of Linlithgow Palace, birthplace of James V and Mary, Queen of Scots, stand on a knoll overlooking the loch. North-east are Blackness Castle and the 17th-century House of Binns. At Bo'ness is Kinneil House, with a museum in its renovated stable block.

WHERE TO EAT
✕✕✕**Champany**
Linlithgow ☎(050683) 4532
Lochside Larder
286 High St, Linlithgow
☎(050684) 7275

GARAGES
Appleyard
Stockbridge, Linlithgow (BL)
☎(0506) 843137

Junction 4

To the south is Torpichen Preceptory which was the principal seat of the Knights of St John.

WHERE TO STAY
★★★**Inchyra Grange**
Grange Rd, Polmont (IH)
☎(0324) 711911

FH Mrs A Hay, **Belsyde House Farm**, Lanark Rd, Linlithgow ☎(0506) 842098
FH Mrs W Erskine **Woodcockdale Farm** Linlithgow ☎(0506) 842088

Junctions 5 and 6 R

Grangemouth's major industry was oil refining long before the discovery of North Sea Oil, since when it has enjoyed a considerable boom.

WHERE TO STAY
★★**Leapark**
130 Bo'ness Rd, Grangemouth
☎(0324) 486733
☆☆**Stakis Park**
Arnott Hill, Camelon Rd, Falkirk (SO) ☎(0324) 28331

WHERE TO EAT
❋✕**Pierre's**
140 Graham's Rd, Falkirk
☎(0324) 35843
✕**Dutch Inn**
Main Rd, Skinflats
☎(03244) 3015
Cladhan
Kemper Ave, Falkirk
☎(0324) 27421

GARAGES
Inch Service Station
Bo'ness Rd, Grangemouth
☎(0324) 484833
Millars
Callendar Rd, Falkirk (Frd)
☎(0324) 21511
Northern
(Johnston Stewart & Sons)
Bainsford, Falkirk (DJ BL)
☎(0324) 22584
Square Deal Motors
Ladysmill, Falkirk (Toy)
☎(0324) 35935

Harthill Services
(Roadchef) 791
Restaurant 06.30–22.00.
HGV cafe. Vending machines. Petrol and diesel 07.00–23.00.
HGV parking. Long-term/overnight parking for caravans £5. Baby-changing. For disabled: toilets, ramp. Footbridge. Credit cards accepted.

Junction 3

Livingston is a fast-growing new town on the River Almond. To the east is the 220-acre Almondell and Calderwood Country Park, situated in and around a picturesque valley.

WHERE TO STAY
♨★★★**Houston House**
Uphall ☎(0506) 853831

GARAGES
Main Service Station
(East of Scotland Tyre Accessories) Livingston
☎(0506) 414191
Redhouse
Main St, Deans, Livingston
☎(0506) 411928

Junction 4

Bathgate is an industrial town, but 2 miles north, on Cairnpapple Hill, is a prehistoric stone circle, recently excavated and laid out. There is also a burial cairn.

WHERE TO STAY
☆☆**Golden Circle**
Blackburn Rd, Bathgate (SW)
☎(0506) 53771

WHERE TO EAT
✕✕✕**Balbairdie**
Bloomfield Pl, Bathgate
☎(0506) 55448

GARAGES
Toll
(I Young Coachworks)
1 Whitburn Rd, Blackburn
☎(0506) 56843
Sam Dornan Car Sales
2–4 Hunters Ln, Whitburn
☎(0501) 40302
Glasgow Rd Autopoint
1 Glasgow Rd, Bathgate
☎(0506) 55152
Nelson
Bathgate Rd, Armadale (BL)
☎(0501) 30343

Junction 5

Hilly country gives rise to many infant rivers around here, including the River Almond. To the south are large areas of coniferous plantations.

Junction 6

The motorway ends here for the time being and the A8 dual carriageway continues until it begins again at junction 8.

WHERE TO STAY
★★**Tudor**
Alexander St, Airdrie (ICA)
☎(02364) 64144
GH **Laurel House**
101 Main St, Chapelhall
☎(02364) 63230

WHERE TO EAT
✕✕**Postillion**
8–10 Anderson St, Airdrie
☎(02364) 67525

GARAGES
Swift Service Station
(L F Hood) Bo'ness Rd,
Chapelhall ☎(0698) 733388
Vicarburn Motors
102–110 Carfin Rd, Newarthill (Sub)
☎(0698) 732684
Coffield Repair Service
53 Carlisle Rd, Auchinlea, Cleland ☎(0698) 860341

Junctions 8 and 9 R

The M73 crosses here – see page 101 for details.

Glasgow's Calderpark Zoo is to the south.

WHERE TO STAY
★★★**Coatbridge**
Glasgow Rd, Coatbridge
☎(0236) 24392

GARAGES
Swinton
218–220 Swinton Rd, Baillieston ☎041–771 4368
A Clark
(West End Motors) 55 Hamilton Rd, Glasgow (BL)
☎041–778 8383

Junction 10

The motorway enters Glasgow's north-eastern outskirts here. A little way north is Provan Hall, a restored 15th-century house set in Auchinlea Park.

WHERE TO STAY
★★**Crow Wood House**
Muirhead (SNB)
☎041–779 3861

WHERE TO EAT
✕✕**La Campágnola**
112 Cumbernauld Rd, Muirhead ☎041–779 3405

GARAGES
Barrachnie Service Station
149 Baillieston Rd, Glasgow
☎041–763 0987

Junction 11

An industrial area on Glasgow's north side.

GARAGES
Jack Fisher Sales and Services
1217 Tollcross Rd, Tollcross
☎041–778 8325

Junction 12

Hogganfield Park, with its loch and island bird sanctuary, is on the north side of the motorway here.

WHERE TO EAT
✕✕**Iram Tandoori**
4 Woodhill Rd, Bishopbriggs
☎041–772 1073

Junction 13

Alexandra Park, on the south side, is another of Glasgow's many green areas.

Junction 14 R

More industrial works and a fruit and vegetable market are around this junction.

Junctions 15 and 16 R

The city centre is just to the south, with its many fine buildings, statues, museums and art galleries. Glasgow Green is a large park by the Clyde containing the People's Palace Museum, a fascinating visual record of the history and life of the city.

WHERE TO STAY
☆☆☆**Stakis Ingram**
Ingram St, Glasgow (SO)
☎041–248 4401

WHERE TO EAT
✕✕**Le Provençal**
21 Royal Exchange Sq, Glasgow ☎041–221 0798
Epicures Bistro
46 West Nile St, Glasgow
☎041–221 7488

GARAGES
C F Taylor Auto Services
33 Charles St, Springburn
(VW Aud) ☎041–552 2517
Craigpark Service Station
432 Alexandra Pde, Glasgow
☎041–554 4058
Applecross Autos
626 Keppochill Rd, Glasgow
☎041–331 1649
H Prosser & Sons
470 Royston Rd, Glasgow
(BL) ☎041–552 4713

Junctions 17 and 18 R

The famous Sauchiehall Street crosses the motorway here and the Museum of the Royal Highland Fusiliers and the Mitchell Library are nearby. To the west is the huge Kelvingrove Park, home of Glasgow's Art Gallery and Museum. At the University, the Hunterian Museum and Art Gallery are outstanding.

WHERE TO STAY
☆☆☆☆**Holiday Inn Glasgow**
Argyle St, Anderston (CHI)
☎041–226 5577
★★★★**Albany**
Bothwell St, Glasgow (THF)
☎041–248 2656
☆☆☆**Crest Glasgow-City**
Argyle St, Anderston (CRH)
☎041–248 2355
★★★★**Stakis Grosvenor**
Grosvenor Ter, Gt Western Rd, Glasgow (SO)
☎041–339 8811
★★**Wickets**
52–54 Fortrose St, Partickhill
☎041–334 9334
GH **Smith's**
963 Sauchiehall St, Glasgow
☎041–339 7674
GH **Kelvin**
15 Buckingham Ter, Hillhead
☎041–339 7143

WHERE TO EAT
✕✕✕**Fountain**
2 Woodside Cres, Glasgow
☎041–332 6396
✕✕**Pendulum**
17 West Princes St, Glasgow
☎041–332 1709
✕✕✕**Ambassador**
19/20 Blythswood Sq, Glasgow ☎041–221 2034
✕**Trattoria Caruso**
313 Hope St, Glasgow
☎041–331 2607
Delta Restaurant
283 Sauchiehall St, Glasgow
☎041–332 3661
Ramana
427 Sauchiehall St, Glasgow
☎041–332 2528/2590
Pizza Park
515 Sauchiehall St, Glasgow
☎041–221 5967

GARAGES
G A Mann & Sons
10–16 Glenfarg St
☎041–332 8865
McGregor & Waddell
36A Buccleuch St, Glasgow
(VW) ☎041–332 1135
A Clarke Automobiles
10 Vinicombe St, Glasgow
(DJ BL Vau Opl)
☎041–334 4761
Ashfield Motors
41 Ashfield St, Glasgow (Frd)
☎041–336 3211

WHERE TO EAT
✕**Bridge Inn**
Ratho ☎031–333 1320

GARAGES
Dougie Miller (Motors)
8A Bridge St, Newbridge
☎031–353 3372
East Calder Service
Camps Ind Est, Kirknewton
☎(0506) 882212

FH Mr and Mrs D R Scott, **Whitecroft Farm**, 7 Raw Holdings, East Calder
☎(0506) 881810
☆☆☆**Royal Scot**
111 Glasgow Rd, Edinburgh
(SW) ☎031–334 9191

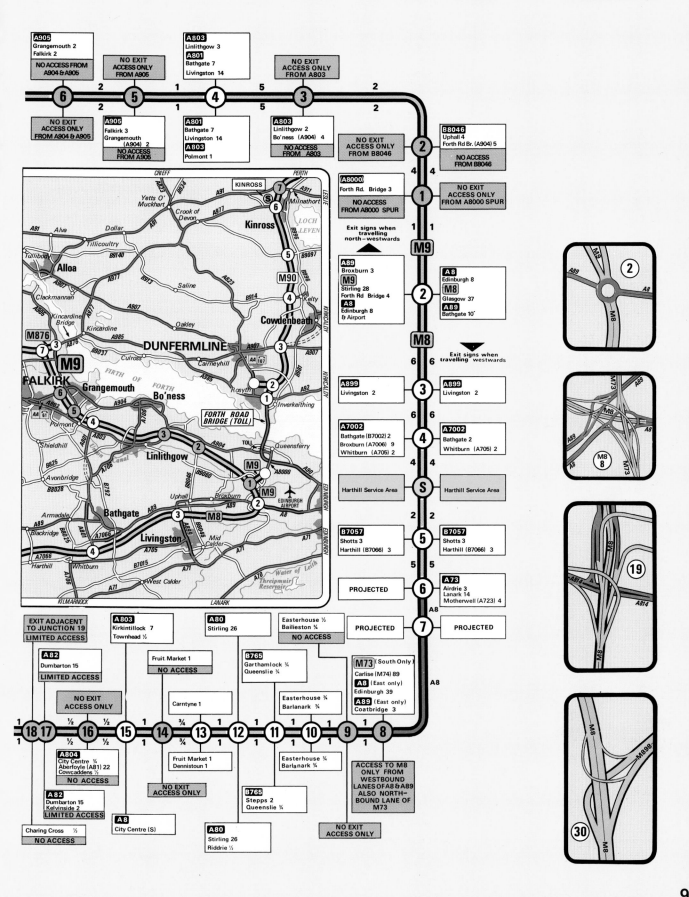

Central Scotland M73 M74 M80

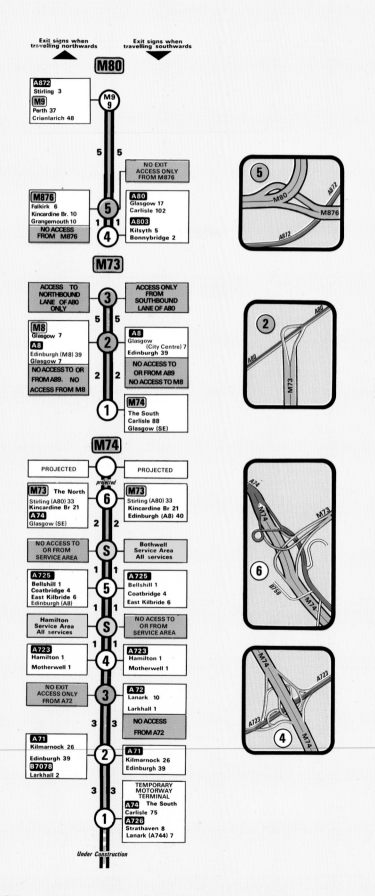

Exit signs when travelling northwards ▲   Exit signs when travelling southwards ▼

M80

A872 Stirling 3 / **M9** Perth 37 Crianlarich 48

M9 9

5 5

NO EXIT ACCESS ONLY FROM M876

M876 Falkirk 6 Kincardine Br. 10 Grangemouth 10
NO ACCESS FROM M876

A80 Glasgow 17 Carlisle 102
A803 Kilsyth 5 Bonnybridge 2

5
1 1
4

M73

ACCESS TO NORTHBOUND LANE OF A80 ONLY

ACCESS ONLY FROM SOUTHBOUND LANE OF A80

3

5 5

M8 Glasgow 7 **A8** Edinburgh (M8) 39 Glasgow 7
NO ACCESS TO OR FROM A89. NO ACCESS FROM M8

A8 Glasgow (City Centre) 7 Edinburgh 39
NO ACCESS TO OR FROM A89 NO ACCESS TO M8

2
2 2

M74 The South Carlisle 88 Glasgow (SE)

1

M74

PROJECTED   ○ projected   PROJECTED

M73 The North Stirling (A80) 33 Kincardine Br 21 **A74** Glasgow (SE)

M73 Stirling (A80) 33 Kincardine Br 21 Edinburgh (A8) 40

6
2 2

NO ACCESS TO OR FROM SERVICE AREA

Bothwell Service Area All services

S

A725 Bellshill 1 Coatbridge 4 East Kilbride 6 Edinburgh (A8)

A725 Bellshill 1 Coatbridge 4 East Kilbride 6

5
1 1

Hamilton Service Area All services

NO ACCESS TO OR FROM SERVICE AREA

S

A723 Hamilton 1 Motherwell 1

A723 Hamilton 1 Motherwell 1

4

NO EXIT ACCESS ONLY FROM A72

A72 Lanark 10 Larkhall 1

3

NO ACCESS FROM A72

3 3

A71 Kilmarnock 26 Edinburgh 39 **B7078** Larkhall 2

A71 Kilmarnock 26 Edinburgh 39

2

3 3

TEMPORARY MOTORWAY TERMINAL **A74** The South Carlisle 75
A726 Strathaven 8 Lanark (A744) 7

1

Under Construction

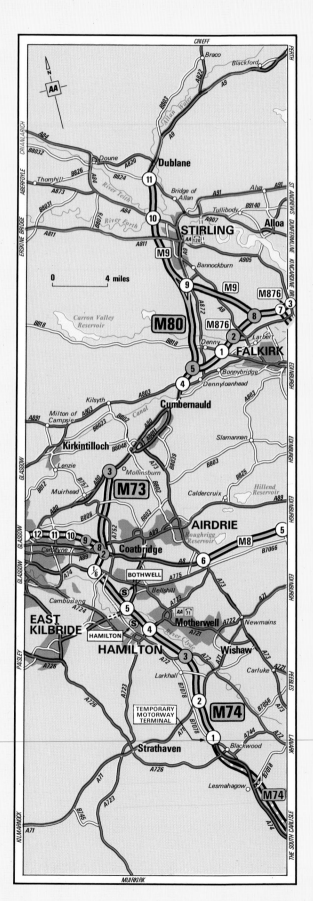

100

M80 M73 M74 Central Scotland
M80

This short motorway forms one side of a triangle with the M9 and the M876, cutting a corner for north-south travellers between Stirling and the south.

The David Livingstone Centre in Blantyre includes the Shuttle Row Museum (above), Livingstone's birthplace and the 'Africa Pavilion'.

and Clyde Canal. To the south is the impressive new town of Cumbernauld.

WHERE TO STAY
FH Mrs J Morton, **Lochend Farm**, Carronbridge
☎(0324) 822778

WHERE TO EAT
Old World Inn
Allanfauld Rd, Cumbernauld
☎(02367) 27509
PS **Garron Valley Forest**
1½m N Garron Bridge
OS57 NS7283

GARAGES
Mill
Bridge St, Bonnybridge (FSO Rel) ☎(0324) 812539
Findlays
24 Barron Hill, The Village, Cumbernauld
☎(02367) 27807

Junction 5 R

The M876 branches off here for the north-east – see page 97 for details.

M9 Junction

The link is complete here where we join the M9 south of Bannockburn – see page 97 for details.

Junction 4

The motorway begins just to the north of the Forth

M73   7 miles

Another part of the system to take Scotland's north-south traffic.

Junction 1

The M74 from the south joins this motorway here – see above right for details. Glasgow's Calderpark Zoo, where birds, mammals and reptiles are housed in spacious new enclosures, is almost beside this junction.

GARAGES
A W L Munro
260 Hamilton Rd, Halfway, Cambuslang
☎041-641 3121
Dechmont
(J May) 84 Hamilton Rd, Cambuslang
☎041-641 3485
D Eadie Cars Ltd
Johnstone Dr, Cambuslang (BL) ☎041-641 3226

Junction 2 R

The M8 continues its journey westwards through Glasgow from here. See page 98 for details.

WHERE TO STAY
FH Mrs M Dunbar,
Braidenhill Farm, Glenmavis
☎(0236) 872319
★★★**Coatbridge**
Glasgow Rd, Coatbridge
☎(0236) 24392

GARAGES
Swinton
218–220 Swinton Rd, Baillieston ☎041-771 4368
A Clark
(West End Motors) 56 Hamilton Rd, Glasgow (BL)
☎041-778 8383
Barrachnie Service Station
149 Baillieston Rd, Glasgow
☎041-763 0987

Junction 3 R

The motorway ends where it joins the A80 between Glasgow and Cumbernauld. A mixture of farmland and industrial sites surrounds the nearby town of Muirhead.

WHERE TO EAT
✕**Neelan Tandoori**
Dalshannon Farm, Condorrat
☎(02367) 20648
Old World Inn
Allanfauld Rd, Cumbernauld
☎(02367) 27509

GARAGES
M P G Motor Company
Unit 5B, Garnkirk Ind Est, Muirhead ☎041-779 1430
Robin Hood Motors
80 Cumbernauld Rd, Muirhead (VW Aud)
☎041-779 1716
Watson Bros
Carbrain Rd, Cumbernauld (Vau Opl Hon)
☎(02367) 25574

M74   11 miles

From the main road leading out of England in the south, this motorway follows the Clyde Valley to the edge of Glasgow.

Junction 1

Another section of this motorway is under construction to the south, but for the time being this temporary junction marks the beginning of the motorway.

WHERE TO STAY
FH McInally, **Dykehead Farm**, Boghead
☎(0555) 892226

GARAGES
Kirkmuir Service Station
(A W and M A Purves) Carlisle Rd, Kirkmuir Hill, Blackwood
☎(0555) 3307

Junction 2

High ground stretches away to the west, with many small tributaries of the River Clyde tumbling down the slopes.

WHERE TO STAY
★★★**Popinjay**
Rosebank (BW)
☎(055586) 441

GARAGES
Old Toll
(T & M Cairns) 128 Main St, Overtown ☎(0698) 74094

Junction 3 R

The River Clyde becomes visible to the east along this stretch of motorway.

WHERE TO STAY
★**Coltness**
Coltness Rd, Wishaw (ICA)
☎(0698) 381616

WHERE TO EAT
Anvil Steakhouse
254 Main St, Wishaw
☎(0698) 375546

GARAGES
Kirk Road Service Station
266–276 Kirk Rd, Wishaw
☎(0698) 73851

Junction 4

A prominent feature of the town of Hamilton, and one which is easily visible from the motorway, is the huge Mausoleum of Alexander, 10th Duke of Hamilton, which was completed in 1854. The town museum is

LOCAL RADIO STATIONS

	Medium Wave		VHF/FM
	Metres	*kHz*	*MHz*
IBA Radio Clyde	261	1152	95.1

Stirling Castle was the residence of Scottish kings as early as the 12th century and it was here in 1543 that the infant Mary, Queen of Scots was crowned.

housed in a late 17th-century inn which has an 18th-century Assembly Hall with musician's gallery.

WHERE TO STAY
★★★**Garrion**
Merry St, Motherwell
☎(0698) 64561

WHERE TO EAT
✕✕**Costa's**
17–21 Campbell St, Hamilton
☎(0698) 283552
✕**Il Frate**
(Friar Tuck) 4 Barrack St, Hamilton ☎(0698) 284379
Pir Mahal
78 Brandon St, Hamilton
☎(0698) 284090

GARAGES
Wellhall
Wellhall Rd, Hamilton (Lad)
☎(0698) 285323

Hamilton Services
(Roadchef) Northbound only ☎(0698) 283446
Restaurant. Shop. Vending machines. Petrol. Diesel. Breakdowns. Repairs. HGV parking. Long-term/overnight parking for caravans £4. Baby-changing. For disabled: toilets, ramp.
Credit cards accepted.

Junction 5

To the east, beside Strathclyde Loch, is the Strathclyde Country Park. Bothwell, to the west, has an impressive ruined 13th-to 15th-century castle and a memorial to Livingstone, both beside the River Clyde.

WHERE TO STAY
★★**Silvertrees**
Silverwells Cres, Bothwell
☎(0698) 852311
★**Hattonrigg**
Hattonrigg Rd, Bellshill (SNB)
☎(0698) 748488

GARAGES
Gopal Motors
350 New Edinburgh Rd, Bellshill (Sko)
☎(0698) 747995
Robb and Allan Motors
Hattonrigg Rd, Bellshill
☎(0698) 747317
Taggarts
5 North Rd, Bellshill (BL)
☎(0698) 748516

Bothwell Services
(Roadchef) Southbound only.
Restaurant. Shop. Vending machines. Petrol. Diesel. Breakdowns. Repairs. HGV parking. Long-term/overnight parking for caravans £4. Baby-changing. For disabled: toilets, ramp.

Junction 6

The M74 ends here where it joins the M73 – see left for details.

M90 Forth Road Bridge–Perth   31 miles

From the Firth of Forth the motorway heads due north to the 'fair city' of Perth, passing the lovely Loch Leven on the way.

Junction 1

The Forth Road Bridge is not of motorway status. That begins here by Inverkeithing, where the Inner Bay provides a yacht marina. To the west are the Naval docks of Rosyth.

WHERE TO STAY
★**Queens**
Church St, Inverkeithing
☎(0383) 413075
GH **Forth Craig**
90 Hope St, Inverkeithing
☎(0383) 418440

GARAGES
Hawes Garage
(G A McTeague) Queensferry
(South) ☎031–331 1796

Junction 2

For six centuries Dunfermline was Scotland's capital and seven Scottish kings are buried there. Dunfermline Abbey is particularly interesting. The millionaire philanthropist, Andrew Carnegie, was born in the town in 1835 and his birthplace is open to the public. He gave Pittencrieff Park to the town and Pittencrieff House now houses a museum.

WHERE TO STAY
☆☆☆**King Malcolm Thistle**
Queensferry Rd, Wester Pitcorthie, Dunfermline (TS)
☎(0383) 722611
★★**Brucefield**
Woodmill Rd, Dunfermline
☎(0383) 722199
★★**City**
18 Bridge St, Dunfermline
☎(0383) 722538
★★★**Keavil House**
Crossford ☎(0383) 736258
★★★**Pitfirrane Arms**
Main St, Crossford
☎(0383) 736132

Junction 3

Another access to Dunfermline, where Queen Margaret (later St Margaret) established the Roman Catholic faith in Scotland.

GARAGES
Clarwood Motors
Henderson St, Kingseat
☎(0383) 734459

Junction 4

The land becomes hilly now, away from the valley of the Firth of Forth. The motorway separates the small town of Kelty from the lovely Blairadam Forest.

WHERE TO STAY
GH **Struan Bank**
74 Perth Rd, Cowdenbeath
☎(0383) 511057

Junction 5

Benarty Hill rises sharply to the east, with a nature reserve on its north side overlooking Loch Leven.

WHERE TO EAT
××**Nivingston House**
Cleish ☎(05775) 216
PS **Loch Leven Picnic Area**
3m SE Kinross OS35 NT1699

Junction 6

The large Loch Leven is to the east beyond Kinross. Loch Leven Castle on an island, was the prison from which Mary, Queen of Scots made her famous escape in 1568. The gardens of Kinross House,

Scone Palace played a major part in Scottish history for well over 1000 years.

between the town and the loch, are also open to the public.

WHERE TO STAY
★★★**Green**
2 The Muirs, Kinross (BW)
☎(0577) 63467
★★**Bridgend**
Kinross ☎(0577) 63413

WHERE TO EAT
××**Kirklands**
High St, Kinross
☎(0577) 63313
××**Windlestrae**
Kinross ☎(0577) 63217

GARAGES
Kirkland's
10 High St, Kinross
☎(0577) 62244
Lochleven Co
High St, Kinross (Frd)
☎(0577) 62424

Kinross Services
(Granada)
☎(0577) 63123
Restaurant 07.00–23.00.
Picnic area. Shop.
Playground. Petrol.
Diesel. Breakdowns & repairs on call. HGV parking. Separate caravan park. Long-term/overnight parking for caravans £3.50.
Baby-changing. For disabled: toilets, ramp.
Credit cards – shop, garage, restaurant.

Junctions 7 and 8 ℞

On the eastern side of Milnathort is Burleigh Castle, a 16th-century tower house with a courtyard enclosure. The motorway continues on its scenic way north through the Ochil Hills.

WHERE TO STAY
★**Thistle**
New Rd, Milnathort
☎(0577) 63222

GARAGES
Stewart & Smart
The Garage, Stirling Rd, Milnathort (Cit)
☎(0577) 62423

Junction 9

Back into the low country of the River Earn Valley with the Firth of Tay to the east.

WHERE TO STAY
★★**Moncrieffe Arms**
Bridge of Earn
☎(0738) 812931
FH Mrs M Fotheringham,
Craighall Farm,
Forgandenny
☎(0738) 812415
★★**Bein Inn**
Glenfarg ☎(05773) 216

Junction 10 ℞

The M85 branches off to cross the Tay for the road to Dundee.

WHERE TO STAY
★★★**Isle of Skye**
Queen's Bridge, Dundee Rd,
Perth (CH) ☎(0738) 24471
★★★**Royal George**
Tay St, Perth (THF)
☎(0738) 24455
★★**Queen's**
105 Leonard St, Perth (TV)
☎(0738) 25471
★★**Salutation**
South St, Perth (AH)
☎(0738) 22166
☆☆**Stakis City Mills**
West Mill St, Perth (SO)
☎(0738) 28281
⊛⊠★★★**Balcraig**
New Scone ☎(0738) 51123
★★★**Murrayshire House**
Montague, Scone
☎(0738) 51171
GH **Clunie**
12 Pitcullen Cres, Perth
☎(0738) 23625
GH **Darroch**
9 Pitcullen Cres, Perth
☎(0738) 36893
GH **Pitcullen**
17 Pitcullen Cres, Perth
☎(0738) 26506

WHERE TO EAT
××**Timothy's**
24 St John St, Perth
☎(0738) 26641
×**Penny Post**
80 George St, Perth
☎(0738) 20867
Pancake Place
10 Charlotte St, Perth
☎(0738) 28077
Kardomah
St John's Sq, Perth
☎(0738) 25093
Pizza Gallery
32–34 Scott St, Perth
☎(0738) 37778
Tudor Restaurant
(Windsor (Perth) Ltd) 38 St
John's St, Perth
☎(0738) 23969
Olde Worlde Inn, City Mill Hotel
West Mill St, Perth
☎(0738) 28281

GARAGES
Heron Rossleigh
Glenearn Rd, Perth (BL DJ LR
Rar) ☎(0738) 20811
Struan Motors
40 South William St, Perth
(Maz) ☎(0738) 33441
Grassicks
50 Leonard St, Perth (BMW
RR Peu) ☎(0738) 25481
T Love & Sons
Citroen House, 60 South St,
Perth (Cit) ☎(0738) 23335

Looking northwards across the Firth of Forth with its famous bridges: road on the left; rail on the right.

Junction 11

The City of Perth still justifies Sir Walter Scott's description of it as 'the most varied and most beautiful' in Scotland. It has two castles – Elcho and Huntingtower – and the historic Fair Maid's House which is now a centre for Scottish Crafts. It has a museum and art gallery as well as the Black Watch Regimental Museum and other places to visit include the Caithness Glass factory. A little way north is Scone Palace where Scottish kings were crowned until 1651. Now home of the Earl of

The motorway passes close to the lovely Loch Leven, seen here from Craigrannoch House.

Mansfield, the Palace contains many fine works of art.

WHERE TO STAY
★★**Lovat**
90–92 Glasgow Rd, Perth
☎(0738) 36555
GH **Clark Kimberley**
57–59 Dunkeld Rd, Perth
☎(0738) 37406
GH **Gables of Perth**
24–26 Dunkeld Rd, Perth
☎(0738) 24717

GARAGES
Letham Service Station
(Buchan Motorcycles)
Rannoch Rd, Perth (Hon Yam
Suz Ves) ☎(0738) 22020
Crieff Rd
(Birrell & Son) Perth
☎(0738) 22067
Grampian Cars
(Rollo) 16 Crieff Rd, Perth
☎(0738) 31515
Strathmore Motors
Arran Rd, North Muirton,
Perth (Vlo) ☎(0738) 22156

LOCAL RADIO STATIONS

	Medium Wave		VHF/FM
	Metres	kHz	MHz
IBA Radio Forth	194	1548	96.8
IBA Radio Tay			
Perth area	189	1584	96.4

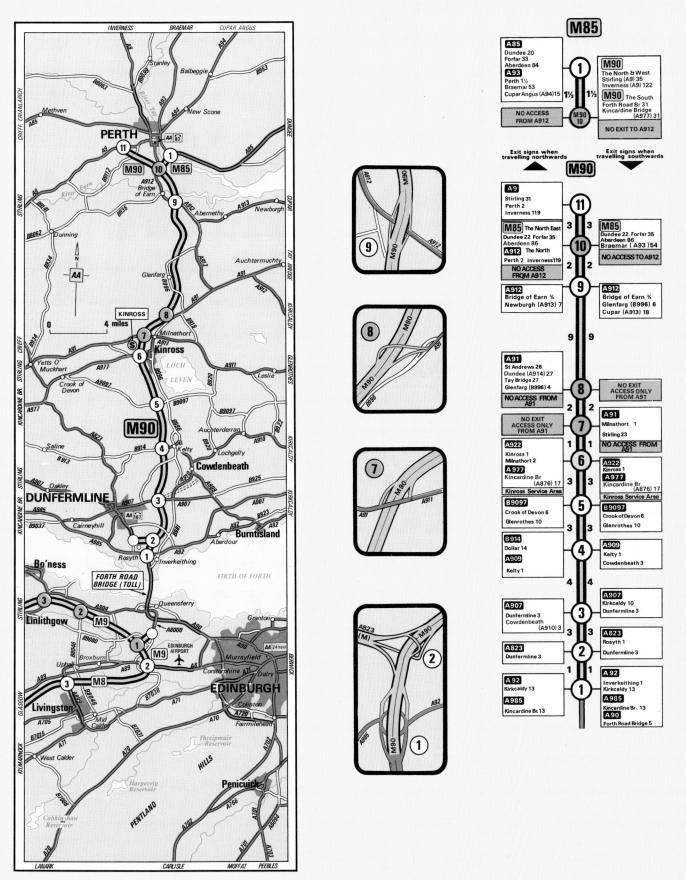

South Mimms - Hatfield - Baldock A1(M)

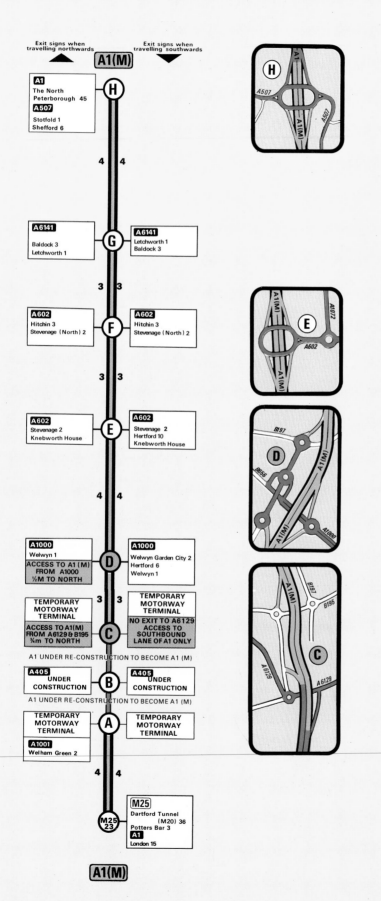

A1(M)

Exit signs when travelling northwards ▲

Exit signs when travelling southwards ▼

A1(M)

H

A1
The North
Peterborough 45
A507
Stotfold 1
Shefford 6

4 | 4

G

A6141
Baldock 3
Letchworth 1

A6141
Letchworth 1
Baldock 3

3 | 3

F

A602
Hitchin 3
Stevenage (North) 2

A602
Hitchin 3
Stevenage (North) 2

3 | 3

E

A602
Stevenage 2
Knebworth House

A602
Stevenage 2
Hertford 10
Knebworth House

4 | 4

D

A1000
Welwyn 1
ACCESS TO A1(M)
FROM A1000
½M TO NORTH

A1000
Welwyn Garden City 2
Hertford 6
Welwyn 1

3 | 3

TEMPORARY
MOTORWAY
TERMINAL
ACCESS TO A1(M)
FROM A6129 & B195
¾m TO NORTH

TEMPORARY
MOTORWAY
TERMINAL
NO EXIT TO A6129
ACCESS TO
SOUTHBOUND
LANE OF A1 ONLY

C

A1 UNDER RE-CONSTRUCTION TO BECOME A1 (M)

A405
UNDER
CONSTRUCTION

A405
UNDER
CONSTRUCTION

B

A1 UNDER RE-CONSTRUCTION TO BECOME A1 (M)

TEMPORARY
MOTORWAY
TERMINAL

TEMPORARY
MOTORWAY
TERMINAL

A

A1001
Welham Green 2

4 | 4

M25
Dartford Tunnel
(M20) 36
Potters Bar 3
A1
London 15

M25 23

A1(M)

H

E

D

C

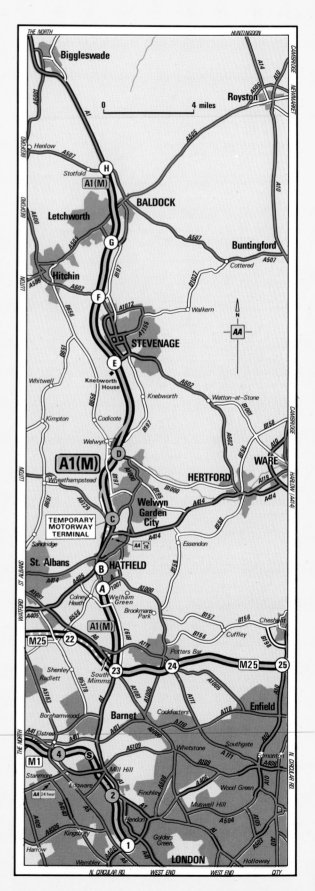

104

Over 1000 acres of Hatfield Forest, including this lovely lake, are in the care of the National Trust. Part of the ancient Royal forests of Essex, it provides excellent walks.

A1(M) South Mimms–Hatfield–Baldock　21 miles open

The A1 assumes motorway status to link these towns to the M25 London Orbital Motorway in the south.

The State Drawing Room of Knebworth House, home of the Lytton family since 1492. Paintings and other treasures cover 500 years of English history.

Junction A

This temporary junction leads to Hatfield where there are a number of interesting old buildings. Hatfield House is the magnificent Jacobean home of the Marquess of Salisbury. In its gardens stands the Old Palace, the surviving wing of the Royal Palace in which Elizabeth I spent much of her childhood.

WHERE TO STAY
☆☆**Comet**
301 St Albans Rd West, Hatfield (EH)
☎(07072) 65411

WHERE TO EAT
Corks and Crumbs
23 Park St, Old Hatfield
☎(07072) 63399

Junction B

This section is under construction and is not expected to open before late 1986.

Junction C

The motorway begins again here on the south-west side of Welwyn Garden City, founded in 1920.

WHERE TO STAY
☆☆☆**Crest**
Homestead Ln, Welwyn Garden City (CRH)
☎(07073) 24336

WHERE TO EAT
PS **Stanborough Park**
E of junction A1/A6129
OS166 TL2211

GARAGES
Brooks & Stratton
Burrowfields, Welwyn Garden City (Ren) ☎(07073) 30678

Junction D

A short distance to the west, at Ayot St Lawrence, is the house where George Bernard Shaw lived. It now belongs to the National Trust and is preserved as a museum.

WHERE TO STAY
☆☆**Clock Motel**
Welwyn ☎(043871) 6911

GARAGES
Godfrey Davis
By-pass Rd, Welwyn (Frd)
☎(043871) 6123
Codicote Motors
High St, Codicote (BL)
☎(0438) 820288

Junction E

Stevenage, a new town which was added to an old market town, is to the north-east. It has a local museum. The Benington Lordship Gardens, overlooking lakes and parkland, include a Victorian folly and an Edwardian terraced garden. This junction also gives direct access to Knebworth House and Country Park. The house has a magnificent Tudor banqueting hall and many art treasures while the country park includes many recreational facilities.

WHERE TO STAY
☆☆☆**Roebuck Inn**
London Rd, Broadwater, Stevenage (THF)
☎(0438) 65444
☆☆☆**Grampain**
The Forum ☎(0438) 350661

Junction F

Hitchin's centre has a number of attractive medieval houses, inns and almshouses. Its museum houses a varied collection including the Regimental Museum of the Hertfordshire Yeomanry.

WHERE TO STAY
☆☆☆**Stevenage Moat House**
High St, Old Town, Stevenage (QM) ☎(0438) 359111
★★★**Blakemore**
Little Wymondley, Hitchin
☎(0438) 55821
GH **Archways**
11 and 21 Hitchin Rd, Stevenage ☎(0438) 66640

GARAGES
Mann Egerton & Co
Queen St, Hitchin (DJ BL)
☎(0462) 50311

Junction G

Letchworth, the world's first garden city, was started in 1903 and its

concept and development are explained in the First Garden City Museum. Another museum contains more ancient Hertfordshire history.

WHERE TO STAY
GH **Butterfield House**
Hitchin St, Baldock
☎(0462) 892701

WHERE TO EAT
Vintage Wine Bar
31 Hitchin St, Baldock
☎(0462) 895400

GARAGES
Swan
(Pugh & Field) Weston
☎(046279) 247
B W Alnutt Motors
20 Blackhorse Rd,
Letchworth ☎(04626) 76657
Vincent Motors
The Green, Letchworth
☎(04626) 74280
Quenby Bros
High St, Baldock (Col)
☎(0462) 893255

The Old Palace of Hatfield, home of the young Elizabeth I before she became queen.

Junction H

The motorway ends here, north of Baldock and the A1 continues northwards To the east Ashwell Village Museum is housed in an early Tudor timber-framed house, formerly a tithe office.

WHERE TO STAY
INN **Three Tuns**
6 High St, Ashwell
☎(046274) 2387

GARAGES
Stotfold Motor Centre
28 Astwick Rd, Stotfold
☎(0462) 730222
Odsey Service Station
(D P Developments) Baldock Rd, Baldock
☎(046274) 2252

The Blakemore Hotel near Hitchen has been considerably extended since it was first built in Georgian style. It stands in large, pleasant grounds of lawns, trees and flower borders.

LOCAL RADIO STATIONS

	Medium Wave		VHF/FM
	Metres	kHz	MHz
BBC Radio London	206	1458	95.1

M11 London–Cambridge   53 miles

The motorway begins just 9 miles north-east of the City of London and ends beyond the beautiful university town of Cambridge. Stansted Airport is passed on the way.

Junction 3

The motorway begins at Wanstead and football fans will find West Ham's ground to the south; Leyton Orient to the south-west. At Stratford is the Passmore Edwards Museum.

WHERE TO STAY
GH **Blenheim House**
2 Blenheim Av, Gants Hill, Ilford ☎01–554 4138
GH **Park**
327 Cranbrook Rd, Ilford ☎01–554 9616
GH **Cranbrook**
24 Coventry Rd, Ilford ☎01–554 6544

WHERE TO EAT
Harts
545 Cranbrook Rd, Gants Hill, Ilford ☎01–554 5000
Spotted Dog
15 Longbridge Rd, Barking ☎01–594 0288

GARAGES
R Powell
374 Eastern Av, Gants Hill, Ilford (BL DJ Tal) ☎01–554 8888
Kaseys Motorcycles
10–14 Woodford Rd, Wanstead Park, E7 (Kaw) ☎01–555 3335
Leawood Service Station
543 Lea Bridge Rd, Leyton, E10 ☎01–539 0095
Markhouse Corner
373/375 Church Rd, Leyton, E10 ☎01–556 1911
New Lighthouse Service Station
124/142 Markhouse Rd, Walthamstow, E17 ☎01–520 0047
Barkingside Motor Co
250–260 Fencepiece Rd, Hainault (Vlo Tal) ☎01–500 0911

Junction 4

To the west is Walthamstow, where the William Morris Gallery shows exhibits of his fabrics, wallpapers and furniture. Here too is the Vestry House Museum of Local History.

WHERE TO STAY
★★★**Prince Regent**
Woodford Bridge ☎01–504 7635
☆☆☆**Woodford Moat House**
Oak Hill, Woodford Green (QM) ☎01–505 4511
GH **Grove Hill**
38 Grove Hill, South Woodford ☎01–989 3344

★★**Roebuck**
North End, Buckhurst Hill (THF) ☎01–505 4636

GARAGES
Gibson Motors
33 George Ln, South Woodford, E18
☎01–989 0262
Hills
536/564 High Rd, Woodford Green (Toy) ☎01–504 9511
W J Wells
Austin House, High Rd, Woodford Green (BL)
☎01–504 0013
Buckhurst Hill
High Rd, Buckhurst Hill (Maz FSO) ☎01–504 7272
Valley Service Station
1 Valley Hill, Loughton ☎01–508 1787

Junction 5

Hainault Forest is to the south-east and Epping Forest to the west. At Chingford is Queen Elizabeth's Hunting Lodge. This picturesque Tudor building houses a museum relating to the wildlife of Epping.

WHERE TO STAY
★★**Roebuck**
North End, Buckhurst Hill (THF) ☎01–505 4636

WHERE TO EAT
×××**Roding**
Market Pl, Abridge ☎(037881) 3030
PS **Hainault Forest**
E and SE of Chigwell Row OS177 TQ4793

GARAGES
Midway
(R J Raven) London Rd, Abridge (037881) 3247
Browns
Browns Corner, High Rd, Loughton (Opl Vau) ☎01–508 6262
Station
(J J Tidd) Station Approach, Theydon Bois ☎(037881) 2451
Wood & Krailing
High Rd, Theydon Bois (Lnc) ☎(037881) 3831
Haven
Stapleford Rd, Stapleford Abbots ☎(04023) 316
Valley Service Station
1 Valley Hill, Loughton ☎01–508 1787
Buckhurst Hill
High Rd, Buckhurst Hill (Maz FSO) ☎01–504 7272
Barkingside Motor Co
250–260 Fencepiece Rd, Hainault (Vlo Tal) ☎01–500 0911

LOCAL RADIO STATIONS

	Medium Wave		VHF/FM
	Metres	kHz	MHz
BBC Radio London	206	1458	94.9
BBC Radio Cambridgeshire	292	1026	96.0
IBA Capital Radio	194	1548	95.8
IBA LBC	261	1152	97.3

Junction 6

This is the junction with the M25 London Orbital Motorway – see pages 74–77 for details.

Junction 7

Harlow was one of the early 'London overspill' towns, but around its modern centre are reminders of its previous history. The Harlow Museum is housed in a Georgian building and nearby is part of a medieval moat. The history of the bicycle is the theme of the Mark Hall Cycle Museum where around 50 machines are on display.

WHERE TO STAY
☆☆☆**Harlow Moat House**
Southern Way, Harlow (QM) ☎(0279) 22441
☆☆**Green Man**
Mulberry Green, Old Harlow (AHT) ☎(0279) 442521

WHERE TO EAT
Beaton's Wine Bar
319 High St, Epping ☎(0378) 72096

GARAGES
Arlington Motor Co
Potter St, Harlow (Vau Opl) ☎(0279) 22391
Kennings Staple Tye
Harlow (BL) ☎(0279) 27541
Kennings First Av
The Stow, Harlow (BL DJ) ☎(0279) 27541
B & G Automotives
Half Moon Ln, High St, Epping ☎(0378) 74753

Junction 8

Bishop's Stortford takes its name from the River Stort and from the fact that the Norman Bishops of London had a castle there. Cecil Rhodes, who founded Rhodesia, was born in the former vicarage, which now houses a museum of his life. Stansted Airport is to the east.

WHERE TO STAY
★★★**Foxley**
Foxley Dr, Stanstead Rd, Bishop's Stortford ☎(0279) 53977
★**Brook House**
Northgate End, Bishop's Stortford ☎(0279) 57892

WHERE TO EAT
××**Bury Lodge**
Bury Lodge Ln, Stanstead ☎(0279) 813345
Swan Restaurant
88 South St, Bishop's Stortford ☎(0279) 52007

GARAGES
Mann Egerton & Co
123 South St, Bishop's Stortford (BL DJ LR) ☎(0279) 58441
J Whalley
London Rd, Bishop's Stortford (Fia Dat) ☎(0279) 54181

Hunts Motor
26 Northgate End, Bishop's Stortford (Peu Tal) ☎(0279) 51388
Concord Motor Services
1 Cambridge Rd, Stansted (Vau Opl) ☎(0279) 813608
Station
(C B Auto Sales) Station Rd, Elsenham ☎(0279) 813127
D Bonney & Sons
34 The Street, Manuden ☎(0279) 813315

Junction 9 R

To the south-east is Audley End House, dating from the 17th century, which contains pictures and furnishings in the state room and a miniature railway in the grounds.

WHERE TO STAY
★★**Crown House**
Great Chesterford ☎(0799) 30515
★★**Saffron**
10–18 High St, Saffron Walden ☎(0799) 22676

WHERE TO EAT
Eight Bells
Bridge St, Saffron Walden ☎(0799) 22790

GARAGES
Cleales
Station Rd, Saffron Walden (Frd) ☎(0799) 23203

Junction 10

Almost beside the junction is Duxford Airfield, a former Battle of Britain fighter station which now houses the Imperial War Museum's collections of military aircraft, armoured fighting vehicles and other large exhibits. There is also a civil aircraft collection which includes Concord 01.

WHERE TO STAY
GH **Highfield House**
55 St Peter's St, Duxford ☎(0223) 832271

WHERE TO EAT
××**Chequers Inn**
Fowlmere ☎(076382) 369

Junction 11

To the east, on the summit of the Gog Magog Hills, is

Wandlebury Ring, the remains of an Iron-age hill fort which was 1000 feet in diameter. About 110 acres of the hills have been protected by the Cambridge Preservation Society.

WHERE TO STAY
GH **Fairways**
143 Cherryhinton Rd, Cambridge ☎(0223) 246063

GARAGES
Harston Motors
High St, Harston (Vlo) ☎(0223) 870123
A J Rayment
28 Woolards Ln, Great Shelford ☎(0223) 843048
Marshalls
Cherryhinton Rd, Cambridge (BL DJ AM) ☎(0223) 249211

Junction 12

Cambridge is a beautiful town which owes its character to the university that dominates its centre. Many of the colleges look out over the River Cam and The Backs, where sweeping lawns and willow trees line the west bank. Among Cambridge's many museums are the Scott Polar Research Institute and a Folk Museum.

WHERE TO STAY
★★★★**Garden House**
Granta Pl, off Mill Ln, Cambridge (PRE) ☎(0223) 63421
★★★★**University Arms**
Regent St, Cambridge (IH) ☎(0223) 351241
★★**Blue Boar**
Trinity St, Cambridge (THF) ☎(0223) 63121
GH **Lensfield**
53 Lensfield Rd, Cambridge ☎(0223) 355017
GH **Helen's**
167–169 Hills Rd, Cambridge ☎(0223) 246465
FH Mrs F Ellis, **Five Gables**, Bucks Ln, Little Eversden ☎(022026) 2236

WHERE TO EAT
×**Peking**
21 Burleigh St, Cambridge ☎(0223) 354755
Eros
25 Petty Cury, Cambridge ☎(0223) 63420
Wilson's Restaurants
14 Trinity St, Cambridge ☎(0223) 356845

Roof Garden and Pentagon
Arts Theatre, 6 St Edwards Passage, Cambridge ☎(0223) 355246
Varsity Restaurant
35 St Andrews St, Cambridge ☎(0223) 56060

GARAGES
Tim Brinton Cars
Hills Rd, Cambridge (Tal Peu) ☎(0223) 213221
Marshall Austin House
Jesus Ln, Cambridge (BL) ☎(0223) 62211
Frank Holland Motors
315–349 Mill Rd, Cambridge (Fia Dat) ☎(0223) 242222
Gilbert Rice
350 Newmarket Rd, Cambridge (Frd) ☎(0223) 315435
Marshalls Car Centre
400 Newmarket Rd, Cambridge (BL DJ) ☎(0223) 65111
Graham Jenkins
(Motorcycles) 29 Cromwell Rd, Cambridge (Yam Puc Lam Ves) ☎(0223) 243074

Junctions 13 and 14 R

On Cambridge's north-western side the motorway ends, with the Fens stretching away to the north.

WHERE TO STAY
★★**Arundel House**
53 Chesterton Rd, Cambridge ☎(0223) 67701
GH **Cambridge Lodge**
139 Huntingdon Rd, Cambridge ☎(0223) 352833
GH **Belle Vue**
33 Chesterton Rd, Cambridge ☎(0223) 351859
GH **Hamilton**
88 Chesterton Rd, Cambridge ☎(0223) 314866
GH **All Seasons**
219 Chesterton Rd, Cambridge ☎(0223) 353386
GH **Suffolk House**
69 Milton Rd, Cambridge ☎(0223) 352016
O **Post House**
Histon (adj A45, 3m north of Cambridge) ☎(022023) 7000
☆☆☆☆**Cunard Cambridgeshire**
Bar Hill ☎(0954) 80555

WHERE TO EAT
××**Three Horseshoes**
High St, Madingley ☎(0954) 210221

A popular pastime – punting on the River Cam between the sweeping lawns of the Backs and the magnificent architecture of the Cambridge colleges.

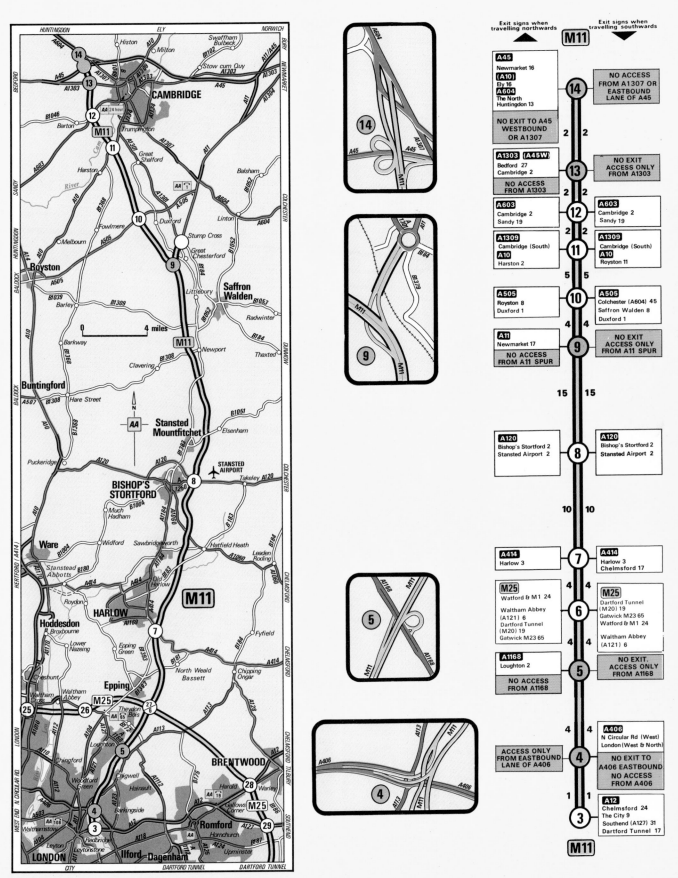

Exit signs when travelling northwards ▲   M11   Exit signs when travelling southwards ▼

A45
Newmarket 16
(A10)
Ely 16
A604
The North
Huntingdon 13

NO EXIT TO A45
WESTBOUND
OR A1307

(14)

NO ACCESS
FROM A1307 OR
EASTBOUND
LANE OF A45

2   2

A1303 **(A45W)**
Bedford 27
Cambridge 2

NO ACCESS
FROM A1303

(13)

NO EXIT
ACCESS ONLY
FROM A1303

2   2

A603
Cambridge 2
Sandy 19

(12)

A603
Cambridge 2
Sandy 19

2   2

A1309
Cambridge (South)
A10
Harston 2

(11)

A1309
Cambridge (South)
A10
Royston 11

5   5

A505
Royston 8
Duxford 1

(10)

A505
Colchester (A604) 45
Saffron Walden 8
Duxford 1

4   4

A11
Newmarket 17

NO ACCESS
FROM A11 SPUR

(9)

NO EXIT
ACCESS ONLY
FROM A11 SPUR

15   15

A120
Bishop's Stortford 2
Stansted Airport 2

(8)

A120
Bishop's Stortford 2
Stansted Airport 2

10   10

A414
Harlow 3

(7)

A414
Harlow 3
Chelmsford 17

4   4

M25
Watford & M1 24

Waltham Abbey
(A121) 6
Dartford Tunnel
(M20) 19
Gatwick M23 65

(6)

M25
Dartford Tunnel
(M20) 19
Gatwick M23 65
Watford & M1 24

Waltham Abbey
(A121) 6

4   4

A1168
Loughton 2

NO ACCESS
FROM A1168

(5)

NO EXIT.
ACCESS ONLY
FROM A1168

4   4

ACCESS ONLY
FROM EASTBOUND
LANE OF A406

(4)

A406
N Circular Rd (West)
London (West & North)

NO EXIT TO
A406 EASTBOUND
NO ACCESS
FROM A406

1   1

(3)

A12
Chelmsford 24
The City 9
Southend (A127) 31
Dartford Tunnel 17

M11

Scotch Corner - Tyneside A1(M)

A1(M)

Exit signs when travelling southwards ▼

A6115 Gateshead 3 Newcastle 4 **A194** South Shields 5 **A1** Morpeth 22 Tyne Tunnel 4	**N** 1 1	
A195 Washington (North) Felling 2	**M** 2 2	**A195** Washington
A182 Washington 1	**L** 1 1	**A182** Washington 1 Houghton-le-Spring 7
A69 Gateshead 5 Newcastle 6 Hexham 28 **NO ACCESS FROM A69**	**K** 1 1	**NO EXIT ACCESS ONLY FROM A69**
Washington-Birtley Service Area All services	**S** ½ ½	Washington-Birtley Service Area All services **A195** Washington (South)
A195 Washington 2 Birtley 3	**J** 1 1	**EXIT TO A195 VIA WASHINGTON SERVICES SLIP ROAD**
A167 Chester-le-Street 1	**H** 6 6	**A183** Sunderland 10 **A167** Chester-le-Street 1
A690 Durham 2 Consett (A691) 15 Sunderland 10	**G** 5 5	**A690** Sunderland 10 Durham 2 Consett (A691) 15
A177 Bowburn 1 Peterlee 11	**F** 6 6	**A177** Bowburn 1 Peterlee 11
A689 Bishop Auckland 7 Hartlepool 15 Teesside	**E** 5 5	**A689** Hartlepool 15 Teesside Bishop Auckland 7
A167 Newton Aycliffe 2 Spennymoor 11	**D** 2 2	**A167** Darlington 4 Newton Aycliffe 2
A68 Bishop Auckland 10 Darlington 3 Corbridge 38	**C** 4 4	**A68** Darlington 3 Corbridge 38
A66(M) Darlington 4 Teesside **NO ACCESS FROM A66(M)**	**B** 3 3	**NO EXIT ACCESS ONLY FROM A66(M)**
	A	**B6275** Barton 1 Piercebridge 5 **A1** The South

Exit signs when travelling northwards ▲

A1(M)

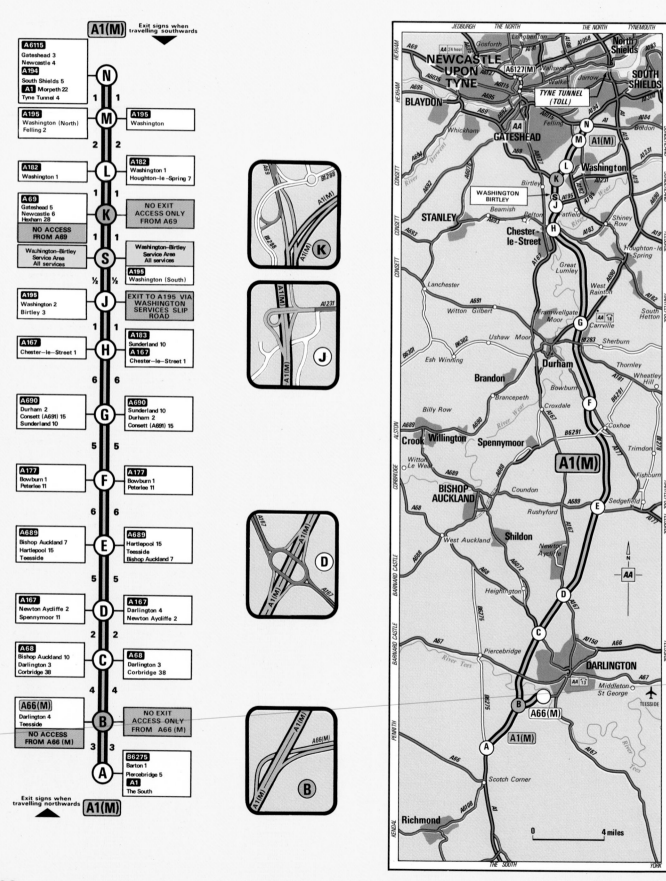

The unusual crown and spire of Newcastle's cathedral overshadow more recent city centre building.

Junction A

The motorway begins just north of Scotch Corner. The Yorkshire Dales are to the south-west.

WHERE TO STAY
★★★**Scotch Corner**
Great North Rd, Scotch Corner (SH) ☎(0748) 2943
INN **Vintage**
Scotch Corner ☎(0748) 4424

WHERE TO EAT
××**Black Bull Inn**
Moulton ☎(032577) 289

GARAGES
Willow Bridge Service Station
Barton ☎(032577) 204
Scotch Corner
Richmond Rd, Scotch Corner ☎(0748) 4831

Junction B R

The A66(M) leads north-east to Darlington. Fame was brought to this town when 'Locomotion No 1' pulled the carriages on the world's first public passenger railway along the line from Shildon to Stockton, passing through Darlington on the way. The original locomotive is among the exhibits at the town's Railway Museum.

WHERE TO STAY
☆☆☆**Blackwell Grange Moat House**
Blackwell Grange, Darlington (QM) ☎(0325) 460111

WHERE TO EAT
×**Bridge Inn**
Stapleton ☎(0325) 50106
××**Bishop's House**
38 Coniscliffe Rd, Darlington ☎(0325) 286666

GARAGES
Cleveland Car Co
Croft Rd, Darlington (BL LR Rar) ☎(0325) 62728
Motor Delivery Co
Grange Rd, Darlington (DJ BL) ☎(0325) 69231

Junction C

North-west at Shildon is another railway museum at the former home of Timothy Hackworth, first manager of the Stockton to Darlington Railway.

WHERE TO STAY
★★★**Kings Head**
Priestgate, Darlington (SW) ☎(0325) 67612
★★**Coachman**
Victoria Rd, Darlington ☎(0325) 286116
GH **Raydale**
Stanhope Rd South, Darlington ☎(0325) 58993

WHERE TO EAT
Taj Mahal Tandoori
192 Northgate, Darlington ☎(0325) 68920
PS **Swan House**
4m NW of Darlington on A68 OS93 NZ2520

GARAGES
G F Bromley & Sons
28–56 Auckland Rd, Darlington ☎(0325) 57228
Whessoe Service Station
(Woodland Rd) Whessoe Rd, Darlington (Opl Vau) ☎(0325) 66044
Sherwoods
Chestnut St, Darlington (Vau Opl) ☎(0325) 66155
Skipper
St Cuthberts Way, Darlington (Frd) ☎(0325) 67581

Junction D

Newton Aycliffe, to the north-west, is an industrial town.

WHERE TO STAY
★★★**Hall Garth**
Coatham Mundeville ☎(0325) 313333

Junction E

At Sedgefield, to the east, there is a racecourse and the Hardwick Hall Country Park.

WHERE TO STAY
★★★**Eden Arms**
Rushyford (SW) ☎(0388) 720541

★★★**Hardwick Hall**
Sedgefield ☎(0740) 20253
★**Crosshill**
1 The Square, Sedgefield ☎(0740) 20153

GARAGES
Turners
Rectory Row, Sedgefield Industrial Estate (Frd) ☎(0388) 817800

Junction F

The hilly country around Bowburn and Coxhoe contains a large number of quarries.

WHERE TO STAY
★★★**Bowburn Hall**
Bowburn ☎(0385) 770311
☆☆☆**Bridge**
Croxdale ☎(0385) 780524

GARAGES
Slake Terrace
West Cornforth ☎(0740) 54761
New Road
(W G Linton) Thornley (BL) ☎(0429) 820302

Junction G

The heart of Durham is almost enclosed in a loop of the River Wear, with its riverside footpath, lawns and trees. At the centre is the superb cathedral, dating from Norman times, and nearby is the castle of similar age.

WHERE TO STAY
★★★**Ramside Hall**
Belmont, Durham ☎(0385) 65282
★★★★**Royal County**
Old Elvet, Durham (SW) ☎(0385) 66821
★★★**Three Tuns**
New Elvet, Durham (SW) ☎(0385) 64326

WHERE TO EAT
×**Squire Trelawny** 80 Front St, Sherburn Village, Durham ☎(0385) 720613
×**Travellers Rest**
72 Claypath, Durham ☎(0385) 65370
Dennhöfers
4 Framwellgate Bridge, Milburngate Centre, Durham ☎(0385) 46777
Mr Toby's Carving Room
Cock o' The North, Farewell Hall, Durham ☎(0385) 43789
The Happy Wanderer
Finchale Rd, Framwellgate Moor, Durham ☎(0385) 64580
Rajpooth Tandoori Restaurant
4 North Rd, Durham ☎(0385) 61496
PS **Finchdale Priory**
1½m N of Durham OS88 NZ2947

GARAGES
Ansa Motors
Carrville (Frd) ☎(0385) 61155
Adams & Gibbon
Claypath, Durham (Vau Opl) ☎(0385) 42511
Fowler & Armstrong
New Elvet, Durham (BL) ☎(0385) 47278
Newton Grange Service Station
Framwellgate Moor ☎(0385) 64036
F Henderson
(Auto Engineers) Ainsley St, Durham ☎(0385) 46319

Ansa Motors
Nevilles Cross, Durham (Frd) ☎(0385) 61155

Junction H

Chester-le-Street was founded by the Romans in the 1st century AD and is thus one of our oldest towns. It is now a business centre. To the west, at Beamish, is the North of England Open Air Museum.

WHERE TO STAY
★★**Lambton Arms**
Front St, Chester-le-Street (SNB) ☎(0385) 883265
★★★**Lumley Castle**
Lumley Castle, Chester-le-Street ☎(0385) 885326

WHERE TO EAT
PS **Holmland Park Transit Picnic Site**
½m from Chester-le-Street on A167 OS88 NZ2753

GARAGES
D Rowe & Co
The Hornet, Chester-le-Street (Frd)
Wadham Stringer
Terminus Rd, Chester-le-Street (BL)
Castle View
Old Penshaw, Houghton-le-Spring ☎(0783) 841061

Junction J

Washington, the place which gave its name to the first president of the USA

and subsequently to its capital, is to the north-east. Washington Old Hall (NT), seat of the Washington family, is open to the public.

> **Washington Services**
> (Granada)
> ☎(091) 4103436
> Restaurant. Shop. Petrol. Diesel. HGV parking. Long-term/overnight parking for caravans £3.50. Baby-changing. For disabled: toilets. Footbridge. Credit cards – shop, garage, restaurant.

Junction K R

The modern, south-Tyneside town of Gateshead boasts impressive athletics facilities and many international events are held there.

Junction L

Another junction for Washington.

WHERE TO STAY
★★★**George Washington**
Stone Cellar Rd, District 12, Washington (BW) ☎091–417 2626
☆☆**Post House**
Washington (THF) ☎091–416 2264

LOCAL RADIO STATIONS

	Medium Wave		VHF/FM
	Metres	*kHz*	*MHz*
BBC Radio Cleveland	194	1548	96.0
BBC Radio Newcastle	206	1458	95.4
IBA Radio Tees	257	1170	95.0
IBA Metro Radio	261	1152	97.0

A sight to make any 'Geordie' homesick – Newcastle's famous bridges carry road, rail and now Metro across the Tyne.

A1(M) Scotch Corner–Tyneside 37½ miles

Once again the long A1 road takes on motorway status as it heads towards the commercial and industrial capital of England's North East.

WHERE TO EAT
××**Ristorante Italia**
580A Durham Rd, Low Fell, Gateshead ☎(0632) 879362
The Griddle
409 Durham Rd, Low Fell, Gateshead ☎(0632) 874530

GARAGES
Inn
(G L Ford & Co) 27 Springwell Ter, Wrekenton ☎(0632) 877041
H Nichols
(Auto Repairs) 610 Durham Rd, Low Fell, Gateshead ☎(0632) 877014

Junction M

Sunderland Airport is to the south-east.

WHERE TO STAY
★★★**Springfield**
Durham Rd, Gateshead (EH) ☎(0632) 774121
☆☆☆☆**Five Bridges**
High West St, Gateshead (SW) ☎(0632) 771105

Junction N

The motorway ends outside the Tyneside conurbation on the south side of the river, but the towns of Newcastle upon Tyne, Sunderland, South Shields, Jarrow, etc are all easily accessible. There are a large number of hotels, restaurants and garages, particularly in Newcastle, but here we list only those closest to the junction.

WHERE TO STAY
☆☆☆**Newcastle Moat House**
Coast Rd, Wallsend (QM) ☎(0632) 628989

GARAGES
Auto Centre
North Rd, Boldon Colliery ☎(0783) 368554
Skewbridge Service Station
Shields Rd, Walkergate, Newcastle upon Tyne ☎(0632) 623301

Index

This index includes places within the text, places on motorway direction panels and the **main** places on the maps. Those marked with an asterisk (★) will only be found on maps.

Acknowledgements

The Automobile Association wishes to thank the following for their help in checking
the information given on motorway service areas

Blue Boar (Motorways) Ltd
Granada Motorway Services Ltd
Kennings Motor Group Ltd
Rank Leisure Ltd
Roadchef Ltd
Sarn Park Services
Trusthouse Forte Service Areas Ltd
Welcome Break Ltd
Westmorland Motorway Services Ltd

Many of the photographs used in *Motorways Made Easy* are the copyright of the
Automobile Association Picture Library. The Automobile Association also wishes to
thank the following for the use of their material.

Blakemore Hotel, Little Wymondley
Cannon Hall, Cawthorne
Crest Hotels Ltd
David Livingstone Centre, Blantyre
Gawthorpe Hall, Padiham
George Hotel, Solihull
Holiday Inn, Heathrow
Red Lion Hotel, Basingstoke
Rose and Crown, Knutsford
Spearpoint Hotel, Ashford